THE LEGEND OF
Atlantis
&
THE SCIENCE OF
Geology

VOLUME 2

The Geology of Greece:
Uniformity or Catastrophe?

JOSEPH O'DONOGHUE

Best Wishes!

Joseph O Donoghue

The Legend of Atlantis and The Science of Geology
Volume 2
The Geology of Greece: Uniformity or Catastrophe?

Print ISBN: 979-8-35092-218-9
eBook ISBN: 979-8-35092-219-6

To Robin and Jordan

And

To My Parents Bridie and Jack

CONTENTS

Social Media and Contact Information

THIS BOOK SERIES IS SUPPORTED BY THE FOLLOWing website and social media channels. The website will undergo continuous development in parallel with the publication of each volume, and additional information will be included. The author is, of course, very interested in the reader's impression and opinion of the books and encourages the reader to visit the site and leave a comment. All responses, comments or questions will be gratefully received and the author will respond to the extent possible, or address categories of questions under a general answer.

Contact the author at: Joe@Atlantis.com

Website: www.atlantisandgeology.com

Facebook: The Legend of Atlantis and the Science of Geology

Reddit: u/atlantisandgeology

X: @atlantis_legend

INTRODUCTION

AS STATED IN THE SERIES INTRODUCTION, THESE first two volumes focus on matters pertaining in any way to the legend of Atlantis. In this second volume, we focus on the other region the Egyptian priest gives us some geological information about, i.e., Greece and the Aegean, which he describes in the context of Atlantis and its destruction. We are, of course, interested to see if the recent geological history of the Greek/Aegean region reflects incremental uniformitarianism, as we're led to believe, or the old priest's rather more dramatic narrative of events.

Having taken a close look at Plato's Atlantis legend itself in volume 1, and from as many angles as could possibly have something to offer, including the geological and astronomical, we should, by now, have a good idea of what it actually says, at least about everything except Greece and the Aegean. We have heard from many of the parties involved, both ancient and modern, including those with one or another axe to grind, most of whom hail from academia, and most of whom condemn the legend as fiction, and its author as an imaginative poet, or worse.

At the same time, we saw that there are still some well-established academics who hesitate to dismiss the legend out of hand, and feel that there may be an historical fact or two lurking in the background. While they are certainly a small minority, these few academics do signify that not all scholars studiously toe the line of orthodoxy. This gives us some modicum of hope, at least.

We spent a good deal of time on the newly created subdiscipline of geomythology and its treatment of Atlantis and other myths. We saw that the approach taken by these academics was little different from any academic anywhere. Most demonstrated disinterest, or perhaps evasion, when it came to giving any myth a proper scientific treatment. This was unsurprising, considering the reluctance of academic geologists to challenge any of the dearly held theories of the uniformitarian establishment, as Berger, for example, made clear. As he suggested, perhaps it's time for some Chinese scientists or physicists to get involved.

In this volume, we will have no hesitation at all in challenging the theories of uniformitarianism, and we will often use physics to do so, while we may meet the odd Chinese scientist also. Since the priest gives us a good deal of geological information about ancient Greece (Athens) we will now use that to conduct a geological "test," shall we say, of Plato's legend, which we could also call his "geological theory." As we saw from Dorothy Vitaliano, in chapter 8 of volume 1, there is little difference between a geological theory and an etiological myth, and hence, Plato's legend is, in essence, an etiological myth concerning the formation of the present-day world.

The priest describes the country in both a general way, as regards its overall size, shape and topography, and also in a more detailed way, pertaining to its soils, sediments, plants, and climate. More importantly, the priest also presents a description of the catastrophe's effects on Greece, simultaneous with the destruction of Atlantis out in the Atlantic.

We have, therefore, quite a lot to work with, especially as compared to some of the legends we examined in volume 1. Still, our examination of those legends showed that some do provide quite a lot of information—much of it unaddressed by academic geomythologists, who clearly hadn't much interest in "elucidating" the nature of the geological events that gave rise to those legends. However, we saw that a simple comparison of one or two legends with the geological evidence suggested that those legends were quite likely valid reports of actual events. This volume is essentially a massively expanded version of that comparison process.

The reader will, of course, understand that much of what I say in this book contradicts the uniformitarian explanation, and much else is considered complete heresy. However, the whole point is to test uniformitarianism, so it can't be helped if the results of the experiment are not to academia's liking—I'm just following the science, as academics themselves say, and I can't help where it leads, as they also say.

We begin with a look at the structure of Greece and the Aegean, both being a single landmass, in reality; it is just that some of it, the Aegean area, is now underwater. We know that the region experiences occasional earthquakes and volcanic eruptions in the present day, and hence is still somewhat active. And here, the first issue arises.

The question is whether this is an actual ongoing process, which has always been occurring at the present rate, or is it merely a newly-initiated process, or, indeed, perhaps it is the fading effect of a much more violent event in the past. This is the first conflict with uniformitarianism, which claims simple and always ongoing continuity. We will see, however, that the Greek evidence indicates past events of a much more violent nature, occurring widely and rapidly, and, indeed, quite recently.

As we move on, we see that past and present geologists interpreted each category of evidence according to the prevailing dogma of the day. Hence, we have an excellent opportunity here to test the evidence against both uniformitarianism and what the priest said of the catastrophe that overwhelmed Greece, as well as Atlantis. This means that everything geologists say about Greece should be reflected in the evidence, and all that evidence should be easily and directly explainable by reference to one or another uniformitarian process currently operating on, or in, Greece, or known, with certainty, to have operated there in the past.

The Great Ice Age is the prime example, and we'll examine that evidence in particular. I must remind the reader here that the Ice Age is not a uniformitarian process; no ice age is ongoing today. And the polar ice caps of today do not represent an ongoing "ice age" at the poles, as some geologists have claimed. Those ice caps merely represent normality.

Rivers, river valleys, and wave action at the coasts are other major subjects of our study, as these are heavily involved in shaping the surface and topography of the country from an erosional point of view. This erosion naturally results in deposition in various other places, filling up hollows, valleys and basins, etc., as well as comprising near-shore deposits such as beaches and estuarine/delta infill.

There are also some rather widespread deposits that are hard to account for. The mountains and valleys, the rivers and lakes, and the soils and sediments of Greece can tell us much about its recent history. Examining all this evidence, we can compare it with what the priest says, and here we also have an early opportunity to examine and assess certain geological theories and principles, to see how they hold up under some critical analysis and field testing.

Greece is particularly famous for its fossil deposits, considered to have accumulated in the last 5 or 10 million years (my), or so. These fossil deposits are mostly composed of the types of fauna that went extinct at the end of the Ice Age, as well as long before it, according to orthodoxy. This will be our first meeting with the megafauna of Greece, and we can be certain that hordes of American Indians at no time swept down from Siberia to run amok among the elephants.

Nevertheless, that Greek megafauna is just as dead as the American one. Furthermore, just as problematic to explain is all the fossil plant life found on Greece, mostly in the form of lignites (brown coal) composed of shredded trees and other detritus, and, indeed, what looks like the remains of entire forests.

The old Egyptian priest discusses Athens and the Acropolis, and the damage suffered during the catastrophe. Here we have the opportunity to directly compare the priest's tale with the evidence on the ground, and we'll find uniformitarianism seriously wanting. The evidence clearly indicates catastrophe and violent flooding, and that evidence is all lying scattered about the vicinity of Athens, in the basins and on the plains and hills surrounding the city.

We can tie much of this evidence directly to the Acropolis and its present condition, noting that this evidence is the same as that found all over Greece. Moreover, the same evidence is found on the islands of the Aegean, including Crete, and we'll see that the same essential cause was also responsible for the evidence found in the basins of the Aegean.

Considering the subject and aim of this series, and not to mention volume 1, it will probably not be a surprise to the reader to find that an analysis of this evidence will leave me with little option but to conclude that major flooding was involved. Having presented my sound scientific reasons why, we'll then see if we can trace the origin and direction of these floods, from the patterns, deposits, and land-forms that evidence has impressed, or left, on the landscape. We will also get an idea of the floods' approximate magnitude.

It may, however, actually be a surprise to the reader to hear that orthodox geology agrees that the Greek region has suffered from a number of major flood events, very much in apparent contradiction of their own uniformitarianism. They even have names for some of them.

We then briefly examine the nature of ancient Greek legends of floods to the extent that we are able, given the lack of serious treatment they have received.

Overall, this volume essentially has three aims. The first is as already stated, to compare the geological evidence found on Greece with what the legend says about Greece and the Acropolis, much as we did for the Native American legends we analyzed in volume 1.

The second aim of the book is to present a detailed picture of the geology of Greece along with the uniformitarian explanation of that evidence. We then analyze both to test the validity of that uni-formitarian explanation, and see if it's up to the challenge.

The third aim is to use this volume as something of a "lesson in geology" for future reference. Therefore, and to forewarn the reader, this volume is heavily geological. but I have made every effort to make it easy to read and understand.

Hence, while the reader may, at times, feel that too much detail, or perhaps too many examples, are provided, there is a reason. And I must note that, in general, such detail will not be entered into in future volumes, apart from certain subjects requiring in-depth analysis.

Whether it achieves these aims remains to be seen, but I can assure the reader that this book presents a very different picture of the science of geology to the one presented by our academics and their institutions.

This will complete our brief study of Atlantis and the geology of Greece. It is intended, and hoped, that the reader will come away with a much better appreciation for the Atlantis legend and catastrophes in general, on the one hand, and perhaps with some skepticism toward uniformitarian geology on the other. And, if nothing else, skepticism is always a good place from which to start.

A Note on Physics

Much mention is made of the laws of physics in this volume. However, I do not cite individual laws, or get into equations or mathematics, restricting any discussion to generalities, or commonly used terms like gravity or force, for example. Since, as noted, we will be discussing flooding quite often, there is one physical law that I must emphasize before we begin. This is Archimedes Law of Buoyancy, known more usually as Archimedes Principle, and is one of the first ones learnt in any high school science class. The law states:

> Any object, totally or partially immersed in a fluid or liquid, is buoyed up by a force equal to the weight of the fluid displaced by the object (Archimedes, c. 246 BCE).

As stated, this law applies whether an object is wholly or partially immersed, and the objects in question in this book are simply the sand, stones, gravel, and boulders that are immersed in water and transported by that water in motion. Hence, if a stone is immersed, it will displace a volume of water equal to the volume of the stone, thereby reducing the weight of the stone as compared to its weight in air.

This means, of course, that the stone is effectively lighter in water, and thus more easily transported. As a matter of fact, most rocks have a density of between 2 and 3, and, since the density of water is 1, most rocks will therefore lose between one half and one third of their weight when immersed. The reader is, therefore, cautioned to bear this principle in mind during discussions of flooding and material transport.

While it is certainly easier to move a boulder in water than it is in air, it is also well to remember that it is still not easy—a three-ton boulder in air is still a two-ton boulder in water, and that is still a hefty boulder. Archimedes principle, by the way, proved out as it has been, is in a different league altogether compared to any of Lyell's so-called principles, as yet unproved, as most of them are.

Further, from a geological point of view, one of the most important aspects of the physics of water is that, like all liquids, water has a coefficient of friction of zero. Without friction, there can be no abrasion, and hence, water by itself cannot erode anything. Thus, there is no such thing as water-erosion, while the term "waterworn" is simply a colloquialism. Water must be carrying abrasive material to effect erosion—by itself, it can only wash away.

Chapter Reference

Archimedes, c. 246 BCE, *On Floating Bodies*.

CHAPTER ONE:

STRUCTURE OF GREECE AND THE AEGEAN

IT IS MUCH EASIER, OBVIOUSLY ENOUGH, TO STUDY the recent geology of Greece and its environs than it is the geology of the 12,500-feet-deep bottom of the Atlantic Ocean, particularly the deeper areas round the Azores platform, that would, with little question, appear to be the area of the Atlantic to which the Egyptian priest was referring.

To put the Greek/Aegean region in context, we recall from volume 1, Chapter 2, where we examined Hesiod's *Theogony*, in which it is stated that Oceanus and Tethys were banished from the Mediterranean, implying that the Mediterranean was closed off from the Atlantic. We also saw that orthodox geology was essentially in agreement, though not with Hesiod's timescale, orthodoxy considering that the closure of the Mediterranean occurred between 12 and 7.5 million years ago, as a result of Africa moving north. Later, at about 5.7 million years ago, the connection with the Mediterranean was restored by the opening of the Pillars of Heracles, which involved a major flood across the Mediterranean, known as the Zanclean Flood.

Hence, according to orthodox geology, Africa collided with Europe and subsequently moved away again. These movements and collisions, therefore, would have had some effect on Greece and the Aegean, which we will examine shortly. The main effects on Eurasia of these large-scale tectonic movements are compression and extension affecting southern areas of the continent, which include Greece and the Aegean, which aren't very far from Africa.

Mainland Greece and the islands are all above water and comprise the visible parts of the Greek continental shelf, which extends eastward beneath the Aegean and is contiguous with the Anatolian landmass, as Lyell himself considered so long ago. Greece and the Aegean, along with Crete, Rhodes, and the rest of the islands, therefore, comprise the area of interest, along with any adjacent areas that may have relevance due to overlap, similarity or contrast. The Aegean at present is a shallow sea, containing many basins and with some deeper troughs in the north and south, and we will examine these also.

Greece has been well-studied geologically and archaeologically, going back to the 19th century, by both Greek and other European scientists, and more recently by international teams as well as oil and gas companies. Most of the hard-rock geology of Greece does not really concern us since we're only interested in the very latest periods, or epochs, these being the Holocene, Pleistocene and earlier Pliocene in order of increasing age. The Aegean Sea has been extensively studied, and much oceanographic work related to structure, topography, sediments and sea level change has been done, which we will examine in due course.

Now, to get back to the Atlantis legend itself and the geological analysis:

Having described the fertility and productivity of the land of ancient Athens, which, according to the priest, far exceeded that of Greece today, or in Solon's time actually, the priest emphasizes that his report is valid or "plausible" and goes on to describe ancient Athens (Greece) as being ". . . like a promontory jutting out from the rest of the continent far into the sea; and all the cup of the sea [obviously the sea-basin] round about it is, as it happens, of a great depth . . ."

The areal extent of the ancient Greece of 11,500 years ago then, would appear to have been much greater, and was, as the priest says, defined by deep water all around. He further comments that great convulsions, or catastrophes, have occurred in the intervening 9,000 years, but in all that time, deluges notwithstanding, there was never any great accumulation of soil that had come down from the mountains.

It is not particularly clear what he means when he says the "soil" instead fell away all round and sank into the depths, or, as Lee (1965) says, "was swept out into the deep ocean." The trouble here is that this statement, "was swept out" sounds reasonable, in that there is a general misapprehension regarding the possibility that something could be "swept" off a continent and into the sea. Sweeping material out to sea from well within a landmass is considerably different to simply sweeping something off a rock by the shore, like a wave sweeping a sunbather away, for example. What *exactly* swept the material all that distance from the Greek interior and off out to sea?

There are few major rivers to speak of in Greece today, and these tend to be in the northern half, near the Balkans. Only two decent-sized rivers flow in the Peloponnese, the Alfios, which enters the Ionian Sea to the west, and the Evrotas, while the Pineios (in Thessaly) flows east into the Aegean, and we will meet others in due course.

In general, all material washed down by rivers in the present day is deposited in floodplains inland, or immediately on meeting the sea.

Some bigger rivers do bring material further out into the ocean, such as the Amazon or Mississippi, for example, but this is only ever fine clays and muds. Waves and sea currents at the seashore also take only the very lightest particles (muds) out into the sea and distribute them.

On the other hand, it takes high energy for rivers to move heavier material (sand, etc.) even as far as the seashore, and then a lot more energy (generally unavailable) for the waves to take it out over the shelf and into the deep ocean, so the meaning of the passage is rather unclear. Since all rivers flow into a mass of essentially static seawater, they slow dramatically on meeting the ocean and drop most of the sediments they are carrying.

In any case, the priest seems to be suggesting that what is left of Greece today, after that greatest deluge of all, is merely the remnant of a much more extensive land area, "... in comparison to what then was, there are remaining..." [depending on which translation] "... in small islets only..." or "... as with little islands..." [or variations of these] "... the bones of the wasted body, as they may be called, all the richer and softer parts of the soil having fallen away, and the mere skeleton of the country being left..."

This reference to a "much more extensive land area" is, of course, essentially what Lyell said of the Aegean area, according to Platt (1889), and as we saw in the prologue to volume 1. Presumably, Lyell took it as evidence of the uniformitarian process of land rising and sinking, and assumed, like Hutton and the rest, that it occurred slowly.

The foregoing passage is something of a puzzle; an enormous flood overrunning the country would, of course, wash away anything loose—soil, stones, rocks—and would sweep them out into the deep ocean. However, the one word that is problematic is the word *soil*, or *earth* in some translations, as in: ". . . the richer and softer parts have fallen away." This seems to suggest that the soil, or earth, was not "washed away" by the flood, as would be expected, but has instead "fallen away," presumably from high ground, with the result that "there are remaining in small islets only the bones of the wasted body."

There seems to be a problem with "soil" from a contextual point of view. The issue is that ". . . the richer and softer parts of the soil having fallen away . . ." doesn't make sense, since one can't really separate good soil from bad, and cause the good only to fall away—or subside, or avalanche—unless it's on hilltops or slopes.

Rich, soft soil tends to be found on the low plains and not on the highlands, and hence it can't very well "fall away." I get the impression that the Greek word for *soil*, as it is used here, could possibly be better translated as "land." Were this interpretation allowed, then the relevant passage would, in my opinion, make a good deal more sense. The rich soft *land* of the plains (along with the plains they rested on) "fell away" into the sea, thereby leaving only the hills jutting above the sea surface as the small islets that constitute ". . . the bones of a wasted body." And this is, more or less, what the evidence says.

While claiming, as I do, that this is a more appropriate translation, the priest also reports deluges, so it's highly likely, if not certain, that there was some "washing away" of the soil also. And, if the perimeter of the promontory that comprised ancient Greece, is, as the priest says, defined by deep water all round, then the areal extent of ancient Greece can easily be defined by tracing the surrounding, and now-submerged, drop-off into deep water. Therefore, the islands of the Aegean would, essentially, form the "bones of this wasted body."

This drop-off, which is really more the inflection point of a slope than a sharp drop, constitutes the beginning of the continental slope. Therefore, by defining the edge of the continental platform on which present-day Greece sits, we get a rough idea of what the former extent of the land used to be, and prior to the "falling away" (subsidence) of the "soil" (land). Figure 1-1 is a 3D depth map of part of the eastern Mediterranean region showing the continental platform of Greece and the Aegean and its continuity with Anatolia.

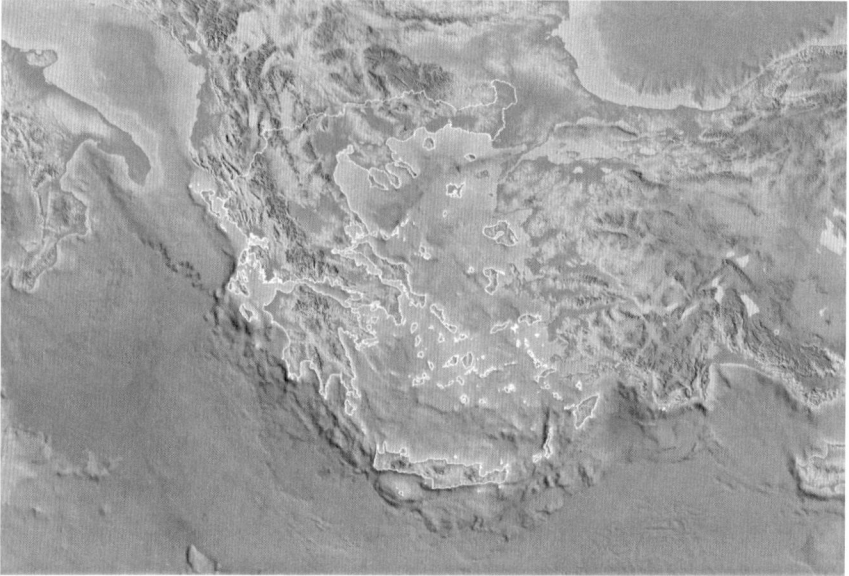

Fig. 1-1: Greece/Asia Minor continental shelf encompassing the Aegean Sea, Crete, and northeast to Rhodes and Türkiye, which would constitute a distinct promontory jutting out from the continental mainland. Image: ID 19376050 Greece outlined. Relief. © Yarr65, Dreamstime.com. Royalty-free Lic.

As can easily be observed, the rough, sloping edge of the continental shelf extends from the northwest, near the boot of Italy, and south along, and close to, the mountainous west coast of Greece and on down to the south of Crete and east and northeast to Anatolia. The deformations and distortions of the seabed of the Mediterranean, visible in the map, strongly suggest that the area has undergone a certain amount of disturbance, with ridges, lineaments and wrinkles, etc., apparent throughout the area of the eastern Mediterranean shown, and with the Aegean region itself looking particularly uneven. We can see the islets remaining above the surface, like the "skeleton" of the country, as the priest says, but it is clear that most of the Aegean area is under water, and we can see the uneven nature of the submarine topography.

THE AEGEAN: Structure and Tectonics

There is no question of the subsidence of this area—as we'll see. Academic geology will entirely agree that the land area of Greece, and especially that of the Aegean Sea region (shown in fig. 1-2) was formerly much more extensive. The whole region was clearly a unified part of the continental landmass of Eurasia, to the point, in fact, that the Aegean Sea did not exist. This is well-attested to by the presence of recent freshwater deposits found throughout the area, indicating the former existence of lakes and rivers where there is now sea.

As we saw in the prologue to volume 1, Greece, and the entire area of the Peloponnese down to Crete, and sweeping across to Türkiye and up at least to the Dardanelles, if not the Black Sea itself, was mostly dry land, and known today as Aegeis (or Aegeida) and shown in the relief map of fig. 1-2, giving a fairly detailed representation of the Aegean and its surroundings.

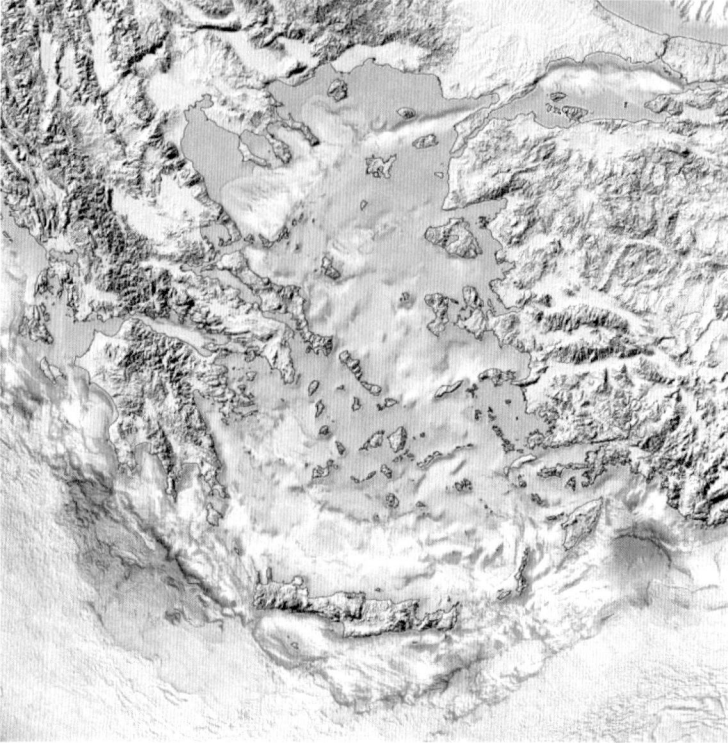

Figure 1-2: Shaded relief map of Greece, Asia Minor and the Aegean region. The Aegean is generally shallow, with submerged plateaus over much of it, and with somewhat deeper basins and trenches in between. Image: Brossolo et al., 2012, Morpho-Bathymetry of the Mediterranean Sea, publication CCGM/CGMW, UNESCO.

While most of the Aegean is relatively shallow, there are some deeper parts in the form of basins and troughs within it. These troughs, without doubt, are due to faults, or cracks in the crust resulting from extensional tectonic movements. The troughs tend to trend in a NW-SE to SW-NE direction, which implies stretching and spreading in a southerly direction as well as differential movement of crustal segments in a generally SW-NE direction as a result of some rotation of Asia Minor, both of which movements resulted in the fracturing of the Aegean tectonic block and its subsidence (a common consequence of such fracturing). The pattern can be seen to continue south of Crete and from there sweep northeast to Anatolia, all of which is in keeping with the southward stretching associated with the reopening of the Pillars of Heracles.

To the northwest, the Peloponnese and nearby peninsulas and islands all trend in a SE direction. They are elevated to the northwest, slope down to the southeast, and appear to continue down into the sea, indicating subsidence, or what's known as tectonic down-warping, in that direction. Like the Azores plateau seen in volume 1, the southern, arcuate edge of the Aegean shelf, though fragmented, transitions rather rapidly to the much deeper water of the surrounding basin of the Mediterranean.

The area at present is still a fairly active tectonic region, as is evidenced by the occasional occurrence of earthquakes and volcanism, which would suggest that it was, relatively recently, the scene of some particularly energetic geological action. There is general agreement on the foregoing, as we'll see, but there is also much debate as to the sequence of events and their timing, and their natures and causes.

Since the big picture here has long been agreed upon, the interpretation of the evidence that sees general agreement is that the tectonic history involves both compression and extension, the former thought to derive from the stress of the African plate impinging on the Eurasian plate. However, that northerly movement is thought to be ongoing and is so shown on tectonic maps (such as fig. 1-4) but is in direct opposition to the extension that is clearly evident over the entire Aegean region.

An early overview of the geological history of the Aegean is given by Eduard Suess in his famous work, *The Face of the Earth*, 1904, once a standard text and still very relevant. Suess deals with the Aegean area on pages 344-345. He begins by declaring: "The most important example of an addition to the sea in very recent times is furnished by the *Aegean* and the *Black sea*." (Suess' italics)

He goes on to say that up until the most recent stage, or period, before the present one, the greater Aegean region was still part of the Eurasian land mass and the Aegean itself contained extensive freshwater lakes. Having studied southern Russia, Kertsch, and the Danube, he concluded that the Black sea region was isolated from the Mediterranean up until recently, this being due to the Aegeis landmass intervening, and over which one or more rivers flowed from the Black Sea to the Mediterranean (Suess, 1904, p. 344). Therefore, freshwater came out through the Dardanelles from the Black Sea, which was most likely at a higher elevation than now in keeping with the higher elevation of Aegeis.

He then cites a paper by Neumayr to the effect that the southern part of the continent first subsided and the Mediterranean encroached as far as Melos, which he, Neumayr, concluded from his finding of recent marine deposits overlying the earlier freshwater deposits of the subaerial Aegeis. This, according to Neumayr, was the beginning of the period of fracturing and the development of the Cyclades Volcanic Arc in the south, along which earthquakes and volcanic eruptions occur at present (Neumayr, 1882, p. 273, ff) and which testify to its recent origin.

The Peloponnese seem to have broken off from the rest of the mainland at the Gulfs of Patras-Corinth, and as far back as 1877, T. Fuchs (1877) determined these gulfs to be major fault troughs and appear to be continuations of the fracture zone related to the volcanic arc, according to Neumayr (Suess, 1904, p. 345) and caused by extension in a S-to-SE direction. Subsequently, according to Suess, final subsidence of Aegeis took place, and did so in very recent time, that is, post-glacial, and so recent he declares that it is even possible these events were witnessed by humans (Suess, 1904, p. 345).

This is evident from the Mediterranean-type molluscs, all but one of currently existing species, that were found in deposits of the Aegean and in the Dardanelles. Oddly, these deposits are found up to forty feet above present sealevel, and a flint knife was said to have been obtained from these beds (Suess, 1904, p. 345). That would confirm human witness and might suggest even more recent uplift subsequent to the earlier subsidence to get those deposits forty feet above sea level, or else some other agent or process was involved.

These deposits add something of a complication, given that sealevel is supposed to have risen 300 to 400 feet in recent times, i.e., after the Ice Age. Hence an uplift of at least 500 feet is needed in order to account for the presence of these sediments at those elevations via uniformitarianist doctrine. That this occurred in the last 12,000 years is implied, which is not acknowledged by anyone, nor is any alternative offered.

There is, of course, an alternative agent capable of depositing large amounts of coarse sediments at these high elevations, but only the priest wants to talk about it. No ordinary river, or high tide or wave action could do this, despite geology's great confidence in the power of such agents, which are entirely inadequate, compared to what's needed here.

Every modern writer on the Aegean reports that its structure is extremely complex, and the tectonic forces and movements that gave rise to it signify a complicated history such that much effort has long been expended in trying to elucidate it, as we will briefly examine. Of course, since Suess' time, the theory of Plate Tectonics has become dominant. Thus, it is compulsory that all modern writers interpret the evidence in terms of that theory, along with maintaining the presumption that all the activity took place over the usual long ages. However, it seems a plate tectonics explanation is not easily forthcoming, which necessitates academia's continued efforts.

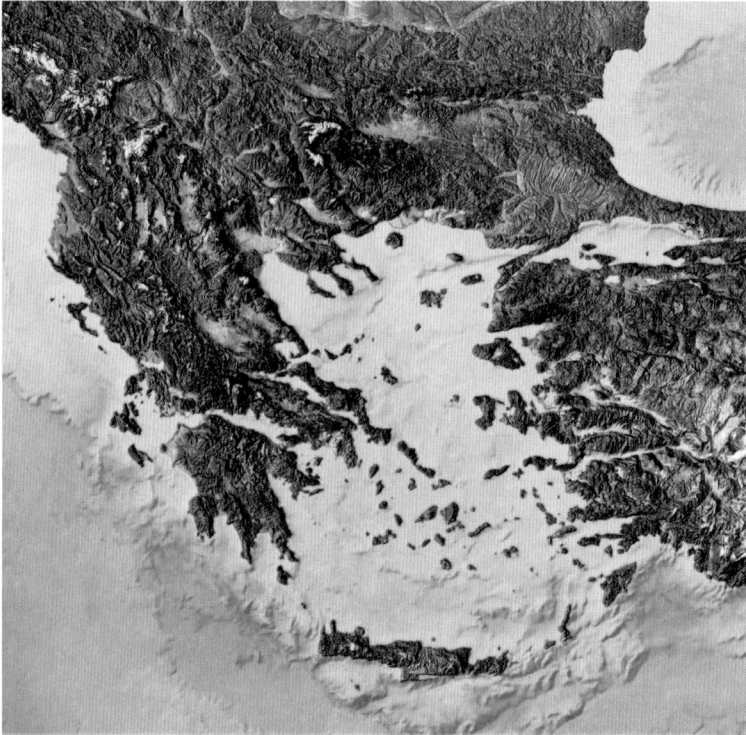

Fig. 1-3: Relief map of Aegeis showing the peninsulas, elongate islands and island chains running in a southeasterly direction before turning and trending to the east and northeast toward Türkiye. The Peloponnese has separated from the Greek mainland, while it, and the other peninsulas and elongate islands, descend in elevation in a southeasterly direction. Image: ID 212152664 © Titoonz, Dreamstime.com. Royalty-free Lic.

This area is considered by modern academic geology to have subsided and uplifted a number of times over the course of geological history, all such interpretations being based on the assumptions that standard theories, tectonics and time-scales are all valid. The initial tectonic uplift was supposedly coincidental with the formation of the Alps, about 30 million years ago.

Subsidence set in at some point; X. Le Pichon & J. Angelier (1981) claim this occurred at about 13 my ago, as opposed to Neumayr (1882) who estimated it at less than a million, i.e., the last era prior to the present. However, it was mostly dry land, according to Suess (1904), until the most recent submergence, supposedly the result of post–Ice Age sealevel rise rather than land subsidence, according to modern writers. At the same time, no one seems to be denying extension, and the evidence for tectonic downwarping is undeniable, such that the Greece/Aegean area is seen to be sloping down to the south.

This Aegeis region, according to conventional geology, fractured due to the collision of the African tectonic plate, moving in a NW direction, with the Eurasian plate, causing some rotation and involving subduction of the Eastern Mediterranean plate under the Aegean plate. Significant extensional, or strike-slip, faulting occurred at right angles, implying that the area was simultaneously stretched apart in an SSW-NNE direction, all of which was accompanied by volcanism (Le Pichon & Angelier, 1981).

These writers are arguing that the entire Aegean landmass has suffered extension in a general southerly direction (arcuate from SW to SE) while simultaneously undergoing compression from this same general direction, in that the African continental plate is presently subducting underneath the Aegean plate in a northward direction while the Aegean plate moves south over it. They claim that the elevated Aegean plate is forcing the subduction zone southward while spreading and thinning in that direction, as indicated in the structural map of fig. 1-4.

This does not make a lot of sense to me, because the authors are suggesting that some force or agent, entirely unidentified, is *currently* stretching and pushing the Aegean plate south *over* the top of the African plate, as *it* pushes north underneath. This, by the way, doesn't make sense to a lot of other people either, because it is obvious that the last thing to have happened here is very distinct southward stretching and subsidence.

Instead, in the brief discussion on page 154 of the same paper, M. F. Osmaston suggests that it may simply be a case of earlier compression followed by later extension, which I consider much more reasonable, considering the foregoing in toto, and the fact that Aegeis was dry land up to recently. Whatever the exact sequence and timescale, Meijer & Wortel (1997, p. 880) claim that present-day conditions date only to the mid-Pleistocene, which is relatively recent, i.e., during glacial times, but actually over a million years ago, which we know to be incorrect, as we've seen from Suess.

Present-day conditions date only to the end of the Ice Age, which is the end of the Pleistocene, a mere 12,000 years ago, not a million or more. Hence, everything was more elevated previous to that time, i.e., during the whole Ice Age. Further, I would not be surprised to find that that also included a much shallower Mediterranean, the floor of which quite likely subsided along with Aegeis immediately to its north, and a shallower and more elevated Black Sea region is also suggested, in keeping with a more elevated and compressed Aegeis. In which case, the foregoing implies that the geography of the greater region was much different 12,000 years ago.

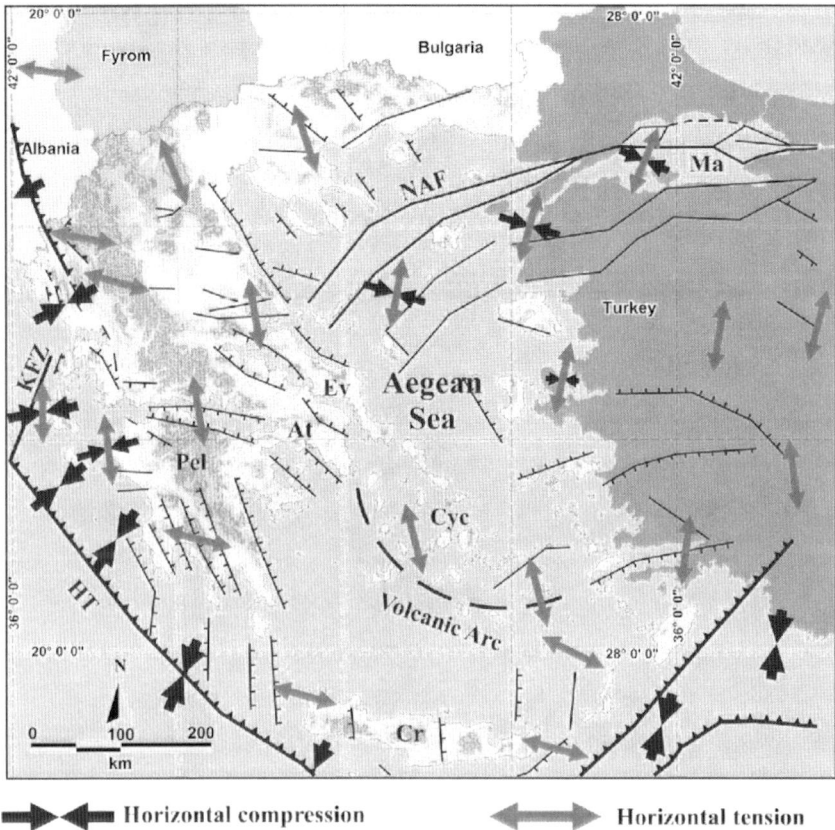

Fig. 1-4: Theorized tectonic setting and stress field of the Aegean and surrounding area. As shown, horizontal compression dominates along the frontal side of the Hellenic Arc (HT), which coincides with the edge of the continental shelf, shown above, while N-S, NE-SW and NW-SE extension prevails in the general Greece-Aegean-Anatolia area north of this. The map, therefore, shows compression and extension in direct opposition to one another, which is highly questionable, at least to me, as is a subduction zone along the Hellenic Arc. Image source: Pavlopoulos et al. 2011, (modified from Hollenstein et al., 2008; Papazachos et al., 1999) see text.

The Plate Tectonics explanation holds that all this movement took place over the course of the last 30 my or so, in one more or less continuous and unidirectional process. However, just about every discussion begins by declaring the region to be extremely complex, and very difficult to explain tectonically, and certainly impossible by a simple collision model.

Even though back and forth motion is not theoretically in keeping with the standard Plate Tectonics Theory (hence Le Pichon & Angelier's "alternative") it has, nevertheless, been widely acknowledged that this best explains the evidence. This would mean, therefore, that Africa collided with Europe, reversed, and is now colliding again, or else no collision is going on at the moment, which is my opinion, based on a lack of any evidence, and the obvious counter-indications.

The current academic "consensus" seems to be compression due to the collision by the northward-moving African plate, westward movement of Anatolia due to pressure from the Arabian plate, and southward extension and subsidence of the Aegean due to (southward) "rollback" (le Pichon & Angelier, 1981) of the African plate, or at least the subduction zone. This means, therefore, that Africa is moving north and subducting under the Aegean plate, while at the same time, the Aegean plate and the subduction zone are both moving south over it.

While these complex movements suggest some kind of Mediterranean merry-go-round, or bumper cars, with everything moving in different directions, it is clear that one can't have it all happen at the same time. Hence two periods of opposite motions and forces must be invoked, and hardly simultaneously. Again, I'm inclined to think that the simple compression followed by extension of M. F. Osmaston is the more correct explanation. When we examine fig. 1-5, it looks like the Aegean has literally been torn to pieces, clearly implying recent extension and not compression.

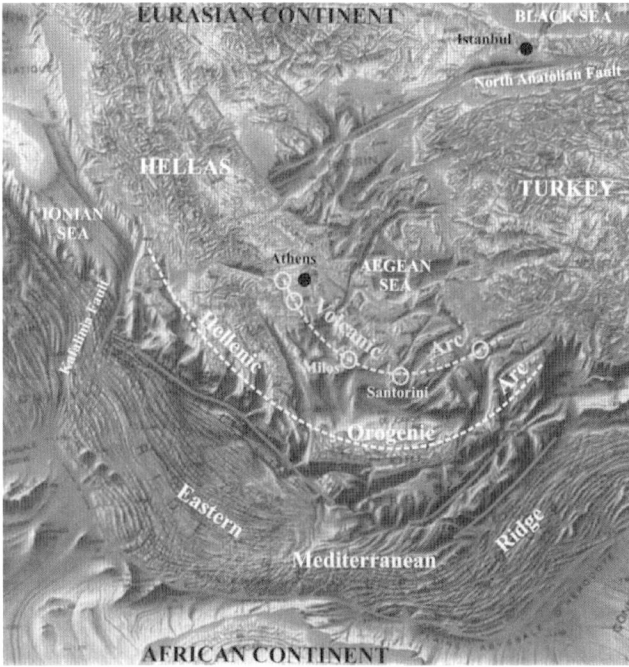

Fig. 1-5: Topographic representation of fig. 1-4, above, with simplified geotectonic data of the Eastern Mediterranean and the Aegean Sea, showing the presumed subduction zone (line with short arrows projecting, near Africa) of the African Plate with the Aegean, the Hellenic Arc and orogenic belt (in white) and the volcanic arc in the southern Aegean Sea. Plain lines are faults with arrows giving known or theorized directions of movement. Source: Dimitris Sakelariou, Hellenic Center for Marine Research, *Science of Tsunami Hazards*, Vol. 30, No. 4, p 257 (2011) image from NOAA Explorer Gallery.

The problem is that, officially, modern geology considers this to have started 30 million years ago or so, and it remains, of course, a typical uniformitarian, ongoing, gradual process. This is a conclusion that is supposedly confirmed by the occasional earthquakes and volcanism that continue right up to the present. Academic geology does know, at the same time, that volcanism and earthquakes were much more extensive, and much more violent, about 12,000 years ago (Huybers & Langmuir, 2009) the period of interest, given the sinking of Atlantis and, indeed, also the time the Aegean Sea came into being.

Hence, the activity of the present day may merely be the fading remnant of the greater activity of that time, rather than a continuation of an ongoing process, as the rate of volcanism and earthquake activity are known positively, from both historical and geological records, to have been declining globally over the last few thousand years, and probably for longer, especially if we take a uniformitarian view of that decline.

Furthermore, the maps above—or any geological map of the region for that matter—clearly show stretching in southwesterly, southerly, and southeasterly directions, which conflicts with the idea that the African plate could be pushing up in this direction, at least at the present time. That, or the Aegean area is no longer stretching to the south and only Africa is moving. I have to consider one or the other, but not both together.

Also, according to the standard plate tectonics theory, subduction of the eastern Mediterranean (African) plate, i.e., pushing north under the Aegean plate, should, theoretically, have caused the Aegean Sea region to crumple up into mountains and rise above that subducting plate, as we see especially with the Andes mountains of South America, where the Pacific plate is supposedly, at present, pushing down under the South American plate and lifting the Andes up.

Therefore, the Aegean area should be high and dry according to the theory. It is not; it is broken up, pulled apart and depressed in a southerly direction, classic evidence of stretching and sinking, and the opposite of what we should find, theoretically at least.

A more recent tectonic study of the Athens city area, carried out by M. Foumelis et al., (2013) designed to measure movements ongoing at the present, used GPS markers to measure such movements in real time. In addition, they also carried out a strain analysis to define the deformation pattern in the area.

Their findings and conclusions serve to at least partly confirm those of just about all earlier workers. Foumelis et al. detected and measured ongoing deformation (strain) in a predominantly NNW-SSE direction. The area continues to be stretched in this direction today, while little compression was observed (negligible in the southern part of the Athens basin) and areas to the east, i.e., in the Aegean Sea area, also show little compression (Foumelis et al. 2013).

Therefore, there appears to be little to support the notion of ongoing northward African movement. It may just simply have occurred in the past, which I would easily accept, given the evidence to be seen.

And, of course, as mentioned earlier, we saw in volume 1 that academic geology considers that Africa collided with Eurasia about 12 my ago, to close off the Tethys in the east, and the Atlantic at the Pillars of Heracles at about 7.5 my ago. The Pillars of Heracles then opened up again about 5 my ago with the Zanclean flood. Therefore, it would appear that a former collision was followed by a pulling-apart, which appears to be what is reflected in the evidence to be seen in the Greek/Aegean region.

What is not clear is which continent moved. Did Africa move north and then south again, or did Eurasia move? Or perhaps both moved. All we can say with reasonable certainty is that there was a collision followed by a partial withdrawal, which is in line with what Hesiod said, and, as we'll see later in this volume, it is also in line with what Pliny the Elder said in his *Historia Naturalis*. It is, of course, also in line with what M. F. Osmaston said.

The complexity of the Aegean Sea region, the lack of agreement on the tectonics involved, and the still-controversial nature of the timing and sequence of events, all serve to show that its history is very difficult for uniformitarian geology to elucidate in any but a very rough way.

A recent paper by Taymaz et al. (2007) provides, from a collection of studies dating from 1979 to 2006, a fairly good summary of the state of knowledge and the theories that have been proffered to account for the Aegean as we find it. As Taymaz et al. state, on page 6, this general region ". . . forms one of the most spectacular and best-studied continental extension regions." Despite all this study, however, little clarity seems to have emerged.

The writers state that extension may be due to (1) slab retreat along the Aegean subduction zone and consequent back-arc extension, (2) collapse of an over-thickened crust, (3) westward escape of Anatolia along its plate boundaries, or (4) differential rates of convergence between NE-directed subduction of the African plate relative to the (hanging wall) Anatolian plate: the latter meaning southwestward movement of Greece relative to Anatolia (Taymaz et al., 2007).

Given the general uncertainty, it is, therefore, rather difficult to have much confidence in any geological narrative of the Aegean region, since academic geology does not seem to have a very clear picture, and there is no "consensus" of which to speak.

While tectonic subsidence, operating in conjunction with one, more, or all of the above-suggested processes, is presently held to account for the subsidence of the Aegean area, this, as usual, is considered to have happened very slowly and long ages ago. Final flooding to more or less present conditions is considered to be due to the approximate 400 ft. rise in sealevel since the Ice Age. It would seem that Eduard Suess and all his evidence have long since been forgotten.

According to a paper on the geographic changes that have occurred in the Aegean region, written by T. van Andel & J. Shackleton (1982), about 18,000 years ago the northern Aegean Sea area was a large coastal plain traversed by many rivers. Broad plains also existed elsewhere, in the Cyclades Plateau, the northern Adriatic, the Gulf of Corinth, and along the Anatolian coast, for example. Many of the islands were connected to the mainland and the Cyclades were mostly joined together.

The land area of Greece was thus much more extensive, just as the old priest claimed, and Lyell himself suggested (vol. 1) and Suess (1904) confirmed. The authors place the main period of sea level rise as occurring between 15,000 and 9,000 years ago. Thus, given that there is general agreement on this point, this time span also brackets the priest's date of about 11,500 years ago, as does Suess (1904) while van Andel and Shackleton (1982) thought, or just simply followed convention, that the Ice Age ended slowly, over thousands of years.

As noted in volume 1, it is now considered to have ended 11,700 years ago, or thereabouts. While there is no agreement whatever on the length of time taken for the ice to melt, it could hardly have still been around 2,500 years after the Ice Age was supposed to have ended. In fact, the date given earlier as 11,711 years ago from the Niels Bohr Institute is derived via analysis of ice cores taken by a team of Dutch scientists on Greenland, engaged in the NordGrip drilling project.

Reporting in the Danish news outlet Politiken, ice-core researcher Jørgen Peder Steffensen declares that the Ice Age ended suddenly, as though a button had been pressed, which means he claims to have detected evidence from the cores attesting to a sudden rise in temperature.

This means that a lot of heat was suddenly available to melt the ice. Another study of ice cores from Greenland was reported by Jeffrey P. Severinghaus, of the Scripps Institute of Oceanography, that suggests that the Ice Age ended in a matter of a few decades and complete melting took no more than a few hundred years (Severinghaus & Brook, 1999), which seems much more reasonable considering the strength of the sun's heat today.

Hence, the 6,000 to 10,000-year melting period of orthodox geology cannot really be considered valid, given such a wide spread, and the close agreement of the ice-core date with that of the priest would suggest an approximate date for the final flooding of Aegeis, whether said flooding was the result of sea level rise or land sinking. Though the reports just given of faster melting come from scientists, only one, Severinghaus, comes from the scientific literature. Said literature most typically holds to the usual orthodoxy of long-term, drip-by-drip melting of the ice sheets over the usual thousands of years, but the evidence of the ice cores is fairly definitive, if valid, and it's also fairly logical.

A much more recent paper on the same topic was published by Simaiakis et al. (2017) who state that sea level rise was accompanied by the *final fragmentation* of the Aegean region, which acknowledges very recent and final tectonic break-up. They follow van Andel and Shackleton in having this occur over a period of 10,000 years (no doubt for the same reasons of orthodoxy, as neither offer any evidence for those extended time periods). According to these authors, the Aegean landmass (Aegeis) lost approximately 70% of its area during this time.

On the other hand, another, even more recent paper on the question of Aegean sea level rise by Louvari et al. (2019) has some sea level rise beginning about 14,700 years ago. However, they invoke "an extremely abrupt marine transgression" at about 11,290 years ago, which, like the Niels Bohr date of 11,711 years ago for the end of the Ice Age, strikes me as just a bit *too* precise. However, it is in line with the priest, at least, as well as with the two ice-core researchers, Pedersen and Severinghaus, referred to above. And, indeed, it is a rare enough thing for any academic to use terms like "extremely abrupt."

While all recent submersion of land areas is conventionally assumed to be due to the post Ice Age rise in sea level, there is no really dependable way to determine whether that submersion is due to sea level rise or land sinking. In all cases, it is a matter of *relative* change of level between land and sea. Actual rise or fall of sea level is called "eustatic change" while the rise or fall of land is called "isostatic change." To tell the difference, of course, one must be able to tell whether the sea level or the land elevation remained static, which one changed, or whether or not both did. And those are not easy things to do, as Theophrastus recognized so long ago, as we saw in volume 1, Chapter 3.

Oddly, in practically all cases where it is obvious that formerly dry land is now below sea level—as in drowned landscapes, estuaries or fjords—the sea level is considered to have risen, while the converse, where the land is now obviously higher relative to sea level—as in raised beaches—the land is considered to have risen. Land sinking under a static sea level is sometimes, though not very often, considered to be the correct interpretation, while sea level actually falling is, naturally enough, never considered, other than for Ice Age initiation and the concomitant water withdrawal necessary for ice production.

The historic emphasis on sea level rise is not simply due to the belief in ice melting; it was formerly thought that land rise, and changes of sea level, were generally localized, as in mountain belts. In fact, it was Eduard Suess, in the same book referenced earlier, *The Face of the Earth*, who first formally stated the eustatic theory, suggesting globally synchronous alternate swings of sea level to explain both sunken and raised land surfaces. Ice ages were, of course, the most obvious way to explain them. However, by the 1930s, continental instability had become undeniable and the concept of isostasy was called on to account for it.

Regardless of Ice Ages and ice melting, etc., the concept of isostasy is entirely theoretical; it involves the idea that continental masses can essentially "bob up and down" on what is theorized to be a highly viscous plastic layer, known as the Asthenosphere, the outer mantle. What speaks against this notion is the fact that the earth's outer, and solid, crust is only 5 to 30 mi. (8 to 50 km) thick, but deep-focus earthquakes occur at depths down to 700 km, as is known from seismic studies.

Also, the two major types of seismic waves, P and S, travel through the mantle, but S waves, which are shear waves, cannot travel through a liquid, like the earth's liquid core for instance. Since S waves do travel through the mantle, the mantle must therefore be solid, given that earthquakes can only occur in rigid materials, according to Newtonian physics.

Since Earthquakes occur due to brittle failure, and, since a plastic or viscous material is obviously not brittle, regardless of how viscous it is, such materials cannot be fractured. I, therefore, have to conclude that up-and-down continental movement must occur by some other means, given that there is no question that it has occurred, as the Aegean, and the rest of the world, demonstrate.

We will discuss the whole issue of solid or viscous mantle in a future volume, but, since the plate tectonics theory requires a ductile mantle, major efforts have gone into explaining deep-focus earthquakes. A good overview of the issue is provided by Arjun Narayanan in the *Berkeley Science Review* of May 21, 2019, wherein he discusses the various suggestions made and the possible causes of the phenomena. However, I personally remain skeptical of a viscous or ductile mantle.

Even though continental instability is well-known, geologists have held more or less solely to the eustatic explanation for general sea level rise and have even produced what's called a Global Eustatic Sea Level Curve, while there is no reliable and permanent datum from which to measure. A good summary of these issues, still relevant today, is given by the well-known Professor Claudio Vita-Finzi of University College London in his book *Recent Earth History* (1973); see Chapter 5, "Ancient Shorelines."

Despite the very obvious difficulties involved in efforts to establish the extent of eustatic versus isostatic sea level changes at any given location, geologists try to isolate the extent of isostatic change to allow for the measurement of eustatic change. They are, of course, certain this eustatic change has occurred due to the melting of all that ice at the end of the Ice Age. Doing so will then enable them to construct a model for the eustatic change only, in terms of extent and duration. In modern times, the computer has greatly aided these efforts since all the components and processes involved can be entered into the program and a general model of sea level rise produced.

The model generated, of course, is entirely dependent on the quality of the data entered, much of which is, in reality, theoretical, estimated, assumed, or consisting of sub-models based on other theories, estimates and assumptions. In which case, the results can hardly be anything other than questionable.

As an example, we have a paper from Kevin Fleming et al. (1998) which claims to further refine the eustatic sea level curve since the height of the Ice Age, the so-called Last Glacial Maximum (LGM). In the abstract, the authors first state that the eustatic component of sea level change provides a measure of the amount of ice (water) transferred between continents and oceans during glacial cycles. They begin by using what they call "realistic" ice distribution and earth models, which is to say they start with a *theoretical* volume of ice, from which they derive a *theoretical eustatic* sea level rise of 125 +/- 5 m.

The earth model is based on the *estimated theoretical* rheology of the planet, which itself is based on the *presumption* of isostasy and a further *assumed* weak, plastic asthenosphere mentioned earlier, rheology being the standard of measure of plasticity (here, the earth's). This is then used to *estimate* the "rebound" from ice unloading and its effect on ocean-water volumes.

In the paper, they provide data on tectonic movements, *assuming* all upward ones are due to isostatic adjustment. Taken together and combined with an *estimated* time taken for melting, they come up with a model for sea level rise taking thousands of years, and which, rather unsurprisingly, hews closely to everyone else's (Fleming et al., 1998).

A brief review of recent literature shows that the question of whether land or sea rose or fell at any location is still as much of a problem to identify as Vita-Finzi said it was 50 years ago. Based on this, it would seem that the accurate history of relative sea level rise or fall at any place on the earth would have to be derived from local evidences only, and thus all such studies and their results apply to a limited area, and no global curve is possible.

Thus far, as I have tried to show, we have no really reliable consensus on the tectonic history of the Aegean Sea, nor is there much agreement on the history of sea level rise and the inundation of the region either. It is, of course, at least mildly significant, but not all that surprising in light of Suess' and others' evidence, that everyone's post-Ice Age sea level rise time span, or window, accommodates the old Egyptian priest's date of approximately 11,500 years ago.

In light of the evidence we have just seen and reports from Suess and others, I am inclined to think that Aegeis stretched and subsided in a southerly direction quite recently, and at the same time that conventional geology has the post-Ice Age sea level rise. I consider that subsidence to be the reason Aegeis is now under water, and not simple sea level rise. I also have to consider that that subsidence was but one small part of the extensive tectonic activity that occurred in the Mediterranean region simultaneously with the subsidence of Atlantis out in the Atlantic. Clarification of this point, though, will have to await our future discussion of the Ice Age.

There is, however, one other aspect of the Aegean that may shed some more light on its recent history, the sediments on its floor. Since the Aegean was only recently submerged, whether by tectonic subsidence, which seems most likely, or sea level rise, which I doubt for various reasons, or a combination of both, it cannot have had a long history of collecting sediment. Thus, we should not expect to find much buildup, and whatever is there should, theoretically at least, consist only of recent, typically marine sediments as found in any similar, shallow sea.

Chapter References

Fleming, K. et al, 1998, Refining the eustatic sea level curve since the Last Glacial Maximum using far- and intermediate-field sites, *Earth Planet. Sci. Lett.* 163, 327–34.

Foumelis, M. et al., 2013, Geodetic evidence for passive control of a major Miocene tectonic boundary on the temporary deformation field of Athens (Greece), *Annals of Geophysics*, Vol. 56, No. 6, Special Issue: Earthquake geology.

Fuchs, T., 1877, Stud. Üb. D. jüngeren tert. Bildungen Greichenlands; *Denkscher Akad. Wiss. Wien*, 1877, XXXVII, 2 Abth., pp 1–42 (cited in Suess, 1904).

Huybers, P. & Langmuir, C., 2008, Feedback between deglaciation, volcanism and atmospheric CO2, *Earth Planet. Sci. Lett.*, Vol. 286, pp 479-491.

Lee, Desmond, 1965, Plato, *Timaeus and Critias,* Penguin Classics, London.

Le Pinchon, X. & Angelier, J., 1981, The Aegean Sea, *Phil. Trans. R. Soc. Lond.*, Vol. 300, pp 357-372.

Louvari, M. et al., 2019, Impact of latest-glacial to Holocene sea level oscillations on central Aegean shelf ecosystems: A benthic foraminiferal palaeoenvironmental assessment of South Evoikos Gulf, Greece, *Journal of Marine Ecosystems*, Vol. 199.

Meijer, P. Th. & Wortel, M. J. R., 1997, Present-day dynamics of the Aegean region: A model analysis of the horizontal patterns of stress and deformation. *Tectonics*, Vol. 16, No. 6, pp 897–880.

Narayan, A., 2019, The mystery of deep earthquakes, Berkeley Science Review, accessed online March 2023.

Neumayr, M., 1882, Zur Geschichte des östl. Mittelmeerbeckens, *Virchow u Holtzendorff's Samml. Gemeinverst. Wiss. Vortr.*, Heft 392, Berlin, 1882.

Simaiakis, S. M. et al., 2017, Geographic changes in the Aegean Sea since the Last Glacial Maximum: Postulating biogeographic effects of sea level rise on islands, *Palaeogeography, Palaeoclimatology, Palaeoecology*, Vol. 471, No. 1, pp 108-119.

Severinghaus, J. P. & Brook, E. J., 1999, Abrupt climate change at the end of the last glacial period inferred from trapped air in polar ice, *Science*, Vol. 286.

Steffensen, J., P., "Ice Age ended suddenly, like a button was pressed" 2008, *Politiken*, Dec. 29, 2008.

Suess, Eduard, 1904, *The Face of the Earth*, (*Antlitz der Erde*).

Taymaz, T. et al., 2007, The geodynamics of the Aegean and Anatolia, *Geol. Soc. Lond.*, Special Publications, 291, 1-1126, p. 6.

Van Andel, T. H. & Shackleton, J. C., 1982, Late Paleolithic and Mesolithic Coastlines of Greece and the Aegean, *Journal of Field Archaeology*, Vol. 9, No. 4, pp 445-454.

Vita-Finzi, Claudio, 1973, *Recent Earth History*, The Macmillan Press Ltd., London.

CHAPTER TWO:

SEDIMENTS AND BASINS OF THE AEGEAN

SEDIMENTARY COVERINGS, AS IN THE NORMAL marine types that collect in all seas and oceans, should be generally thin, since, as noted, the Aegean is a very young sea. This goes especially for the more elevated areas of the seabed that formed the uplands of the former land surface of Aegeis, and where the water is now shallow. Being elevated, they should not have much relict sediment from the earlier times when Aegeis was sub-aerial, as hilltops are, obviously, not typical places where sediments accumulate. As described and shown earlier in various figures, the structure of the Aegean Seabed is one of plateaus, ridges, basins, and troughs (see figs. 1-2 and 1-5 previous, and 2-1 below).

The plains and basins that existed prior to the submergence, which, according to Suess (1904, p. 344) and Louvari et al. (2019), contained extensive lakes and rivers, can be expected to have a good deal more sediment. At the same time, the later marine material should be clearly distinguishable from, and overlying, the normal lacustrine, river, and floodplain deposits that would have collected on a previously subaerial Aegeis.

Sediments expected to be found are the normal ones found in seas and oceans, i.e., the remains of pelagic microfauna, possibly some muds, wind-blown terrestrial dust from the general area or fine silt from the Sahara, cosmic dust, and, but only very close to the coasts, terrestrial material washed into the sea by rain and rivers.

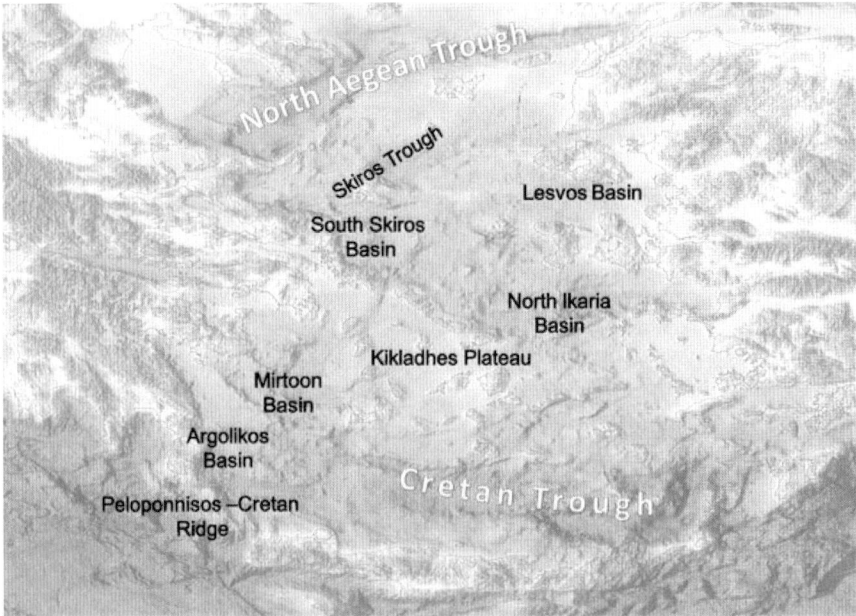

Fig. 2-1: The Aegean area, showing features such as basins, troughs, plateaus, and coastal shallows. Image from: Deep-sea Bio-Diversity in the Aegean Sea. Gönülal & Dalyan (2016). Image: Intechopen CC BY 3.0.

The material derived from land may be moved, along with any coastal erosion products, by surface waves and sea currents and redeposited close to shore. Of this sediment, only the finest mud will be distributed anyway widely about the sea. Sediments on the ocean floor are, therefore, generally of two types, those derived from the ocean or sea itself, in the form of debris or detritus representing the remains of oceanic animals or plants, i.e., biogenic, or those derived from land and referred to as terrigenous. As far as the Aegean goes, then, only these two types of sediments should be found covering the sea floor, with a minor amount of cosmic or Saharan dust.

What interests us in particular about these sediments is the type, location, thickness, and, most especially, origin. The main questions are the nature of the sediments and how they got to where they are now, given that, according to standard uniformitarian geology, only the kinds of processes and agents acting in the area at present were acting in the area in the past, Ice Age or no Ice Age.

Admittedly, those agents were obviously acting on a much different landscape, one mostly above sea level, and thus rain, rivers, frost-action, wind, earthquakes and volcanic eruptions and a limited amount of marine action in the south, much as we see acting today the world over, must, theoretically at least, be responsible for everything found at present on the floor of the Aegean and off its continental slope to the south.

Many workers have studied the Aegean in recent years, and there are many reports on the nature of the sea floor surface sediments in various parts of the sea. Serafim Poulos of the University of Athens gives a synthesis of these reports (Poulos, 2009).

The results are exactly as one would expect—coarse-grained terrigenous material in the near-coastal zones, especially in the north, brought in by rivers draining northern Greece and the Balkans for the most part, as well as the mountains of Greece to the west, and the coast of Anatolia to the east. Biogenic material predominates over the central and southern parts of the sea, with fine mud/clay dispersed more generally all over the sea floor by circulating sea currents. Some dust, presumably Saharan, is present, and some fine material is brought in through the Dardanelles and the straits in the south near Crete.

The foregoing inventory of Aegean Sea floor sediments briefly describes the results of the standard uniformitarian processes that have operated in the Aegean region during the latest, or most recent, period in the earth's history. Obviously, this is since the Aegean region was submerged, and therefore, presumably that is, since the end of the Ice Age, regardless of how that submergence was accomplished.

Because of the changes relating to the latter, we should expect to see some difference in the types of sediments comprising the top layer and those immediately underlying it. Poulos, however, does not provide the thickness of the top layer of sediments or any information on those lying beneath.

As described earlier, the Aegean Sea floor comprises a series of plateaus and troughs or basins. The large troughs, such as the North Aegean and Cretan, are tectonic in nature and can be considered more as underwater rift valleys than basins. Large basins generally trend in various directions related to the extension suffered, and are often deep and steep-sided. Known as grabens (German for "grave"), they are due to subsidence resulting from tectonic extension. The plateaus of the Aegean Sea floor incorporate shallow basins and some north-south grabens due to east-west extension. The following diagrams give a general idea of the structure underlying, and producing, the horst and graben topography.

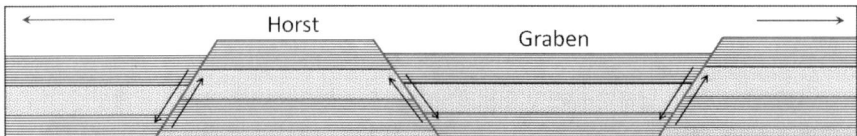

Fig. 2-2: Simplified horst and graben block diagram. Image USGS.

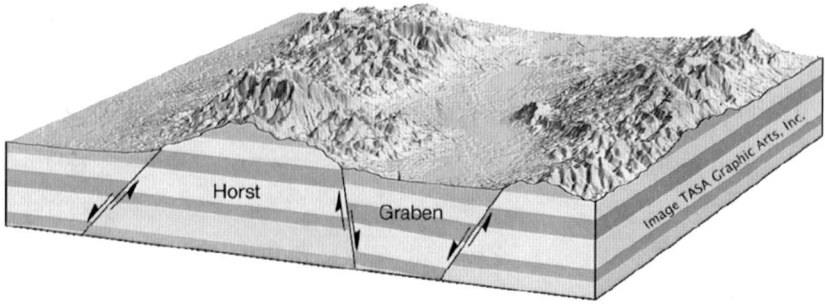

Fig. 2-3: More realistic perspective representation of horst and graben structure and resulting topography, similar to that found in the Greece of today, and widespread over the country. Image: Wood et al., 2015 http://geology.isu.edu/Alamo/Devonian/basin_range.php.

In the Aegean, therefore, the grabens are where sediments will collect so long as some agent operates to deliver sediment into them from the land, while whatever marine (biologic) materials or mud, etc., that are in suspension in the seawater drift down into them. These basins are by no means regularly shaped and may be curved, tapering, or otherwise distorted. The faults bounding them, the result of their formation, are not necessarily neat, singular planes inclined at the same angle throughout. Those boundary faults may instead be multiple (fault arrays) and lie at varying angles, while the floors may also not be horizontal, as is often shown in block diagrams, but may be uneven, broken up and lying at varying angles.

Many seismic profiles have been taken and the sea floor has been drilled in many areas, especially in the basins during oil exploration programs and scientific investigations. If we take the sea-bed-surface sediments to be of very recent age, i.e., Holocene period of the last 11,700 years, then the underlying sediments will be of Pleistocene and earlier Pliocene age, which are the supposed Ice Age and earlier sediments.

However, the literature, irritatingly, refers to all the sediments, including the most recent, as Plio-Quaternary sediments,
for some reason making no distinction among pre–Ice Age, Ice
Age and post–Ice Age sediment types, all of which should be
distinctly different. In any case, the total time period assumed
for the deposition of all this material is approximately 5 million
years, according to stratigraphic theory. A fairly comprehensive
survey of Mediterranean Plio-Quaternary sediment thicknesses,
derived from seismic surveys and drill cores, was published by M.
Gennesseaux et al., in 1998.

In the section dealing with the Aegean (pp 277-279) the authors
report Plio-Quaternary sediment thickness in the Thasos basin
section of the North Aegean Trough of 3,000 meters, or almost
10,000 ft., and overlying older sediments of a further 3,000 m. To
the northwest in the Gulf of Thermaïkos, a drill core showed 3,500
m of "often coarse, clastic sediments," of which 1,800 m (5,500 ft.)
were Plio-Quaternary.

In the Sporades basin section of the North Aegean Trough, the
drill reached 5,000 m, of which the top 2,000 m (6,500 ft.) of Plio-
Quaternary sediments were of the same coarse, clastic type. These
clastic sediments are now well offshore, in deep water. Logic, or at
least its close relative, would suggest that they were derived from a
landmass to the north or west, or both, or else they came from the
east through the Dardanelles.

Wherever they came from, it necessarily requires fast and
powerful currents of water to transport "coarse, clastic sediments"
one or two hundred miles from the seashore to where we now find
them, and where was that water flowing if the Aegean area was above
sea level during this whole period of the Ice Age? One would need
multiple, large, powerful, high-speed, and persistently flooding rivers

flowing off the Greek and Türkish mountains, to do this work, or else, perhaps, a couple of overwhelming deluges.

In the central region, comprising the shallow Aegean Plateau, the basins are filled with deposits more than 1,000 m (3,000 ft.) thick, but the authors do not tell us their nature. (Gennesseaux et al., 1998) The authors do, however, consider them to be extensions of the Anatolian coastal valleys. One larger basin, the North Ikarian, is also shown in fig. 2-1, and it looks a good deal like the subsidence basin typical of the whole Aegean, when compared with those in fig. 2-4.

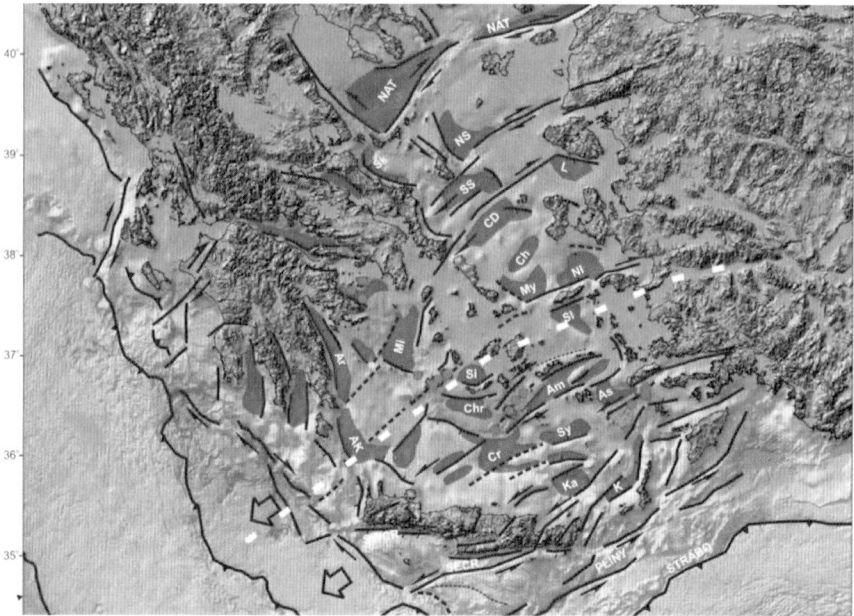

Fig. 2-4: Simplified tectonic map of the Aegean Region with major faults and fault zones, and the many basins of the area. Abbreviations of basins: NAT: North Aegean Trough, NS: North Skyros, SS: South Skyros, Sk: Skopelos, L: Lesvos, CD: Cavo d' Oro, My: Mykonos, NI: North Ikaria, SI: South Ikaria, Mi: Mirtoon, Ar: Argolikos, Si: Sikinos, Chr: Christiana, Am: Amorgos, As: Astypalea, Sy: Syrna, Cr: Cretan, Ak: Antikythera, Ka; Kamilonissi, K: Karpathos. Image: Sakellariou et al., 2013.

The authors note, however, that over the general Cyclades Plateau (Aegean Plateau) area, the sedimentary cover is irregular, as one might expect from a region that was once above sea level, was topographically uneven, and contained numerous lakes. This Aegean region was therefore once a plateau with a somewhat undulating surface.

Many of the basins referred to in the text are shown in the bathymetric/topographic map in fig. 2-4, particularly those around the Central Aegean Plateau, or present-day Cyclades Plateau. As a matter of interest, fig. 2-5 shows a seismic profile across the Amorgos Basin (a double basin) and the Astypalea Basin south of it, shown in fig. 2-4, and located in the south-central region of the Cyclades Plateau, trending NE – SW, in line with most faults.

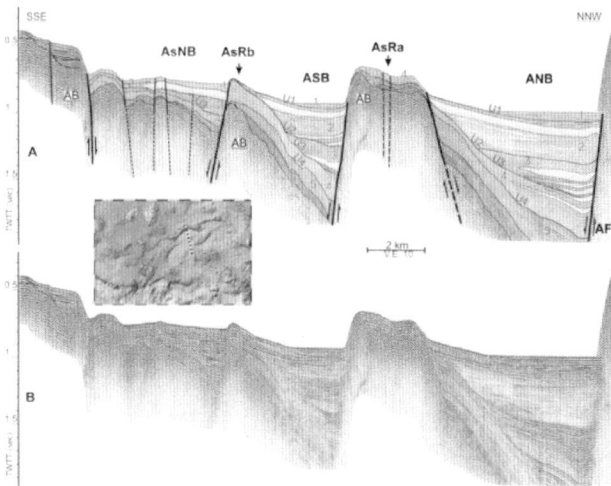

Fig. 2-5: Seismic profile across the central part of the Amorgos North Basin, the eastern part of the Amorgos South Basin and the western part of the Astypalea North Basin, shown as Am and As, in fig. 2-4. Note that the vertical scale of the diagram, like all seismic profiles, is in seconds and therefore does not reflect the thickness, but the *time* taken for the seismic wave to pass through the rocks and sediments. Image: Tsampouraki-Kraounaki et al., 2021.

The profile clearly shows the horst and graben structure of the basin complex, along with tilting of the subsided blocks. The authors identify six sedimentary units (1-6) and tell us there are some differences among sediment types but that they are all generally similar in each basin, based on their interpretation of the seismic data. Some chaotic deposits are inferred from the seismic results, which the authors interpret as volcanic in origin, i.e., pyroclastic flows from a nearby volcano.

They also identify four unconformities (breaks in the sequence) shown in the seismic section as U1 to U4, indicating erosional episodes between different depositional phases, and possible tectonic activity.

U1 to U3 were formed since the basins subsided while U4, common to both the basins and ridges, clearly predates the breakup of the Cyclades Plateau (Tsampouraki-Kraounaki et al., 2021). Sedimentary formations (Units 1-3) that lie above U3, therefore, represent relatively recent deposition, with breaks (U1 and U2) and erosion of the top surfaces of each of the lower units. All sediments thicken to the north, indicating, to the authors, that that is the direction from which the infill originated.

As is normal with this and other seismic profiles, as well as cross-section diagrams in general, the vertical scale is highly exaggerated, usually by at least 3 or 4 times and sometimes much more, as here where the authors give it as 10:1. As noted, the vertical scale is in time units and, therefore, layer thickness, as shown, represents the amount of time taken for the seismic wave to pass through each layer. At the same time, the vertical scale is separately exaggerated in the interests of clarity, to distinguish the individual layers. However, such exaggeration always gives a deceptive impression of greater thickness, or higher relief.

While the basins shown have three sedimentary formations in common, all restricted to the basins themselves and lying above unconformity U3, the authors tell us that much the same situation prevails in the surrounding basins, many of which were included in this study program. All of the basins studied show much the same profiles with the same sedimentary wedges filling the basins the same way, and separated by equivalent unconformities.

The top unit, here shown as forming the seabed, is not the most recent; the most recent sediment consists of the usual thin layer of biogenic material mixed with fine mud, volcanic dust, etc., amounting to only a few centimeters or inches, that is continuously being deposited at present, and has been since the deposition of Unit 1 was complete. Unit 1, therefore, is the last sedimentary deposit laid down prior to present conditions being established, which, as we know, was at the end of the Ice Age a mere 12,000 years ago.

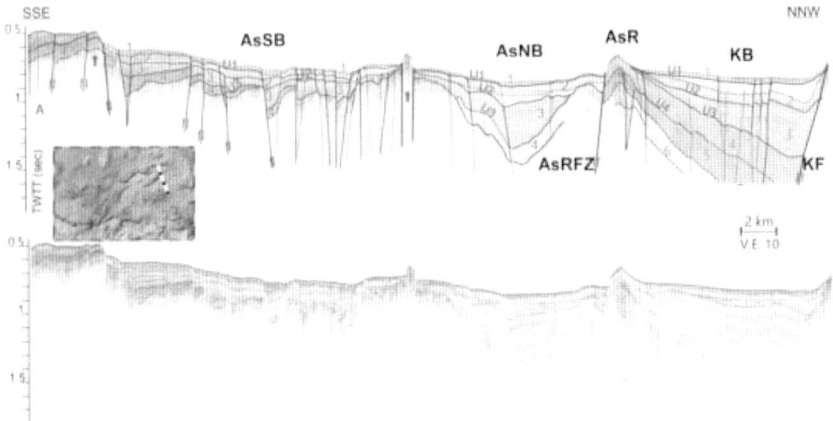

Fig. 2-5a: Profile through the Kinairos, Astypalea South and Astypalea North Basins. The sedimentary sequences and unconformities are seen to be common to all basins in the profiles shown. Image: Tsampouraki-Kraounaki et al., 2021.

As mentioned, Unit 4, underlying Unconformity 3, is more or less continuous both in the basins and on the ridges separating them, indicating that Unit 4 predates the formation of the horst and graben structure, after which Units 3 to 1 were sequentially deposited. And this is where the problem arises. As can be seen, the volume of fill in any basin greatly exceeds the volume of the top of any ridge that projects above the surface of the sedimentary fill, even if we project the fault planes upward to their intersection and thereby increase the volumes of the ridges.

The vertical exaggeration, of course, makes this a little difficult to picture on the diagrams shown. The volume of the infill, therefore, means that erosion of the tops of the horsts cannot have been the origin of the sediment, southward into each basin. Instead, it must have come from further north, beyond the basins, and indeed, from outside the general Cyclades Plateau itself.

The authors assert that the infilling occurred synchronously with the subsidence of the basins, in which case, one has to ask how the sediment was swept into one basin, thence over the horst and into the next basin, and again over the next horst and so on. There being no water movement to speak of at the bottom of the sea, the filling must have occurred when the basins were sub-aerial. In which case, where did all the needed water come from and how was it travelling fast enough across a wide flattish plateau, and over multiple ridges, to transport such an enormously large quantity of sediment sufficient to fill up any number of basins.

They are all filled in a similar way, and with generally similar sediment types. The agent responsible, therefore, must have operated over an extensive area, and it left all the basins with nice, neat layers, as shown in fig. 2-5. Only a generally acting and powerful flood, operating on a large and broad scale, could perform this kind of work. And that conclusion is drawn directly from the evidence, and logic, given that we are required to explain why basins over such a broad area are all in much the same condition. Geology, as we'll see later in this volume, is finally acknowledging this, at least up to a point.

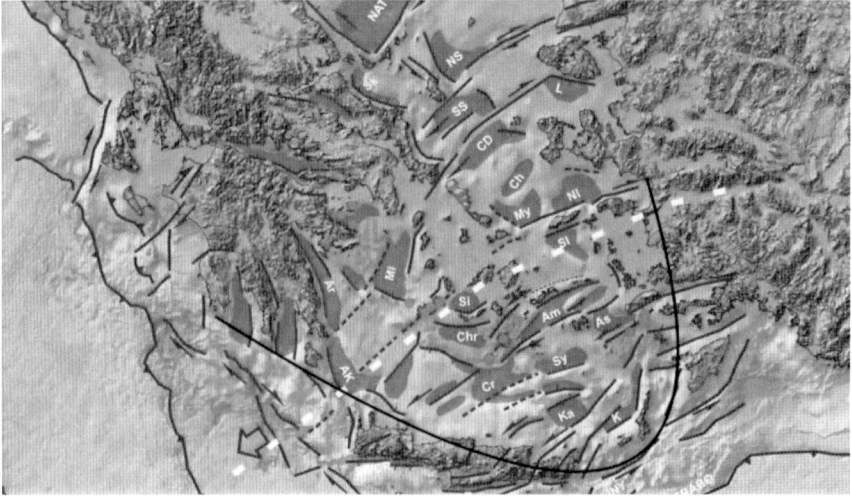

Fig. 2-6: Central section of fig. 2-4 showing many of the basins mentioned (smaller ones not shown). As can be seen, the Peloponnese, Attica, and Euboea are to the northwest of the Cyclades Plateau, while the shallow region, encompassing the basins, extends to the southeast as outlined by the black line.

If we imagine the seismic cross-sections more realistically as much flatter, the actual plateau surface and the generally shallow topography of the area preclude the possibility of large fast-flowing rivers carrying large amounts of sedimentary material. The cross-sections indicate a sediment source in the Northwest, i.e., the Greek mainland, and the evidence seems to suggest that by some means this material was swept over the Cyclades Plateau and deposited in the basins, and this action was repeated three times, whether in quick succession or separated in time.

The basin cross-sections show three sedimentary formations in each one, bounded above and below by unconformities that the authors report as erosional in origin. Whether the subsidence of the basins was slow or rapid, something swept material into them, of a nature and at a rate sufficient to erode the underlying material of Unit 4. Subsequent to this, a second something repeated the same process, eroding the top surface of Unit 3 and depositing another formation, Unit 2. This action was repeated a third time to deposit Unit 1, prior to present conditions being established, and represented by the thin top layer measuring only a few inches, or cms, thick, which could not represent a very long time.

In a short paper on the palaeogeographic evolution of the Cyclades Islands, i.e., those parts of the Aegean Platform that remained above sea level after the final subsidence of Aegeis, K. Gaki-Papanantassiou et al. (2010a) focus mostly on tectonic action and eustatic sea level changes. However, in the paper, they also mention the "peneplains" of Greece, peneplains being extensive areas of flattish land. They note that Greece has quite a number of such plains, of different ages and at different elevations, interpreted as due to tectonic movements (Riedl, 2004).

Here, no isostatic movement or adjustment is inferred, the uplift being caused by movements of the various tectonic blocks involved, as converging plates typically cause uplift due to compression.

The peneplains are stepped as though one formed first, was lifted up, subsequent erosion formed a lower one, and it too was elevated. Each peneplain is thought to be formed by normal erosion, and finally, if erosion continues to its ultimate conclusion, the so-called peneplain will be reduced down to sea level, which is known as "base-level." This is essentially a throwback to early uniformitarian theory of how the land surface gets worn down.

To digress for a brief word on peneplains. The term *peneplain* was coined by William M. Davis of Harvard University in 1899, and it meant a "near-plain" formed by the uncompleted erosional work of a river, specifically. It did not, initially, address other types of plains, or near-plains, formed by other processes, such as coastal uplift, marine erosion, glacial activity, or other uniformitarian process. In actuality, the term *peneplain* referred to an uneven surface, generally flattish or gently sloping and with an uneven or undulating topography.

Later, due to its usefulness as a descriptive term, it became applied to any erosion surface fitting that description. It is also the type of plain long thought to be formed by marine erosion and is much theorized about by Lyell and many other geologists of the 19th century. They held that the erosion occurred during phases of uplift or subsidence as the land moved through the zone of wave action. Peneplains were one component of what Davis, in his 1899 paper, called the "Geographical Cycle," more usually referred to as the "cycle of erosion," which Davis specifically linked to fluvial erosion and uplift (Davis, 1899).

In the meantime, the peneplain, as J. D. Phillips (2002) tells us, is a "venerable concept in geomorphology, geology and geography. Yet despite more than a century of effort, no convincing example of a contemporary peneplain has been identified, and the identification of relict peneplains is uncertain and controversial."

This "venerable" concept of the peneplain has its roots, as noted, in the 19th century, in the concept of "marine planation" effected by the sea while a land area was subsiding or uplifting. In fact, this process is a major component of uniformitarian geology, as it is believed to be one of the most effective agents of erosion, and is credited with forming much of the topography of the earth.

Without being specific as to particular agency or process, the term "peneplain" now means a gently sloping or level plain of low-relief formed by erosion of some kind. Despite a lack of supporting evidence, and the fact, stated by Phillips (2002), that none are seen to be in the process of formation anywhere today, this great concept yet lives on, along with all the rest of those concepts, "principles" and theories that serve as answers in place of actual knowledge.

As we will see, W. M. Davis and his Geographical Cycle were widely embraced for the first 50 years of the 20th century, until such time as the very serious difficulties with it became apparent (this is usually code for no supporting evidence). Skepticism grew, challenges became widespread, and the concept was widely rejected by the 1950s. However, with the rise to prominence of Plate Tectonics Theory in the 1960s and 70s, Davis's Geographical Cycle was resurrected and modified such that when combined with other geomorphological systems, many somewhat questionable themselves, it served an explanatory function in certain situations (despite being demonstrably invalid).

Exactly what kind of "explanatory function" an invalid concept could have is quite beyond me. Davis himself provided no examples of his supposed end-stage: a peneplain at sea level. In his book *Geographical Essays*, 1909, in reference to his Geographical Cycle, he boldly states: ". . . the scheme of the cycle is not meant to include any actual examples at all, because it is by intention a scheme of the imagination and not a matter of observation . . ." (Davis, 1909, p. 281).

While fantasy is all well and good, where appropriate, the world the science of geology deals with is *certainly* a matter of observation—not some "scheme of the imagination."

As Phillips (2002) says, a peneplain is a ". . . logical outcome during a period of prolonged tectonic stability . . ." which, as everyone knows, is supposedly the normal state of affairs on this planet; everything happens slowly and everything takes forever. We are given to believe that plate tectonics and all those other, related, tectonic processes also happen extremely slowly. Any point or period at any time in earth's history, therefore, should be a point or period of apparent and ongoing stability, with change happening too slowly to be perceptible. Volcanic eruptions, earthquakes and floods, etc., are all part of that stability.

If current conditions as we see them around us have obtained for lengthy periods of geological time, and if all those uniformitarian eroding processes we also see going on have actually been working as we're told they do, then the world should be littered with peneplains.

Surprisingly, it isn't, as J. D. Phillips tells us on p. 226: "However, extensive, low-relief erosional surfaces worn to near sea level are at best rare, and by some reckonings, nonexistent" (i.e., imaginary, just as Davis said,). What, then, are we to make of the multiple Greek "peneplains" identified for us here?

The authors, Gaki-Papanantassiou et al. (2010) go on to report that these old peneplains derive from chemical weathering and erosion by supposed sheet-floods, and the younger ones have suffered destruction recently by claimed linear erosion, i.e., by rivers (fluvial erosion). In either case, they have been worn down by moving water, which is to say moving water necessarily carrying debris, sand, etc., since water by itself cannot erode hard, consolidated rock or anything else, but can only *wash away* loose, broken and friable materials, like unconsolidated volcanic ash, for example, shattered rock, sand, mud, etc. Therefore, the water must have been moving quickly to wash away and carry the eroding material.

One must wonder how water could flow quickly and powerfully across a flattish, low-relief "peneplain," given the requirement of a slope to get the water moving, and a steep one to get it moving quickly enough to pick up or wash away debris and other materials. It must then sweep this abrading material over the surface of the peneplain to accomplish all that erosion. And then it would have to do this repeatedly, in what appears to be the complete absence of the required conditions.

Given that, how is the water to achieve this on the gigantic peneplain that is the Aegean Plateau, in the absence of the hills, valleys, and steep slopes needed both to collect large amounts of rainwater and cause it to flow with the high velocity required to pick up debris and sweep it over the surface, so as to render the Aegean in the condition we now find it? Any rain will collect and flow in valleys, not over the general land surface.

To the east of the Aegean Plateau lies the Ikarian basin (NI and SI on fig. 2-6). A study by Lykousis et al. (1995) on the sediments in these basins—that is, the sediments also studied later by Gennesseaux et al. (1998)—identified four sedimentary series they describe as "oblique progradation delta sequences." This means that, according to these writers, the sediments in question are (or were) building out into the basin, having been washed off the land surface and transported into the basin by a river building a delta into it.

They mention "turbiditic deposits" which, of course, are deposits supposedly resulting from (entirely theoretical) "turbidity currents" and are identifiable by their particular sedimentary profiles. The series are typically composed of coarse clastic material at the base, which grades up through progressively finer material, with variations, to end at fine mud at the top. This is known as a classic "Bouma Sequence," after the man who first defined them.

Turbidity currents are thought to ultimately result from the buildup of river sediment on a submarine shelf-slope to an unstable condition, whereupon a disturbance causes a submarine slump, or a landslide, that then (somehow or another) rushes off down the shelf slope, which is typically no more than 4 or 5 degrees, and does this through static, high-pressure seawater, and then flows way off out across the level seabed.

Turbidity currents have been the standard go-to explanation for the presence of *coarse clastic* sediments on the seabed since the early 1950s, shortly after such sediments were first found far from shore. They have been wholeheartedly embraced ever since as explaining the otherwise inexplicable existence of these sediments in various places, and over huge areas, where, according to uniformitarianism, these sediments just should not be. That "turbidity currents" have never been observed, or otherwise detected, despite 70 years of searching, does not seem to matter.

In a similar way to a theoretical "turbidity current," a declining flood will, of course, drop the very heaviest, coarsest material first, followed by progressively lighter material as the flow speed and water volume decline, while the next and subsequent floods will reproduce much the same profile leading to a repetition of the profile stacked one above the other, just like stacked Bouma sequences. The standard, or ideal, Bouma sequence has various features such as fine laminations and or ripples, among others, and is rather particular.

The ideal, or complete, sequence, it should be noted, is rarely, *if ever*, found in the wild and it can be difficult to distinguish a supposed Bouma sequence from a simple run-of-the-mill flood deposit. We will discuss the whole question of turbidity currents later, as they have a major bearing on the interpretation of all the, shall we say, "anomalous" evidence we'll find at the bottom of the Atlantic.

In the Amorgos and associated basins (see fig. 2-4) along the southern margin of the Cyclades Plateau, Nomikou et al. (2008) report that the basins are all bounded by active (as usual) faults that display the usual extension in a NW-SE direction. They report the Quaternary sediment thicknesses at up to 700 m (2,300 ft.).

Further to the west, regarding the Mirtoon basin and adjacent areas, Anastasakis et al. (2006) also report both turbidite deposits and fan-deltas, the turbidites originating from inferred rivers flowing from the north, Attica region, and/or off the Aegean Plateau when it was above sea level (Anastasakis et al., 2006). Thus, we see that the inferred origin of the sediments of Tsampouraki-Kraounaki et al. (2012) is indeed the Greek mainland, even if the means of transport are debatable.

In the south, in the Cretan Trough, the deposits are of the order of 1,000 m (3,300 ft.) and again consist of the same Plio-Quaternary deposits. In a paper on the geology of Mediterranean Sea basins in general, Biju-Duval et al. (1974) agree with other workers on the Aegean region, and also state that the basins comprising the Hellenic Trough and those around Crete and Rhodes are all filled with recent sediments up to 1,500 m (4,800 ft.) thick (Biju-Duval et al., 1974, p. 713), wherever they came from. There are few rivers flowing off Crete or Rhodes, and 1,500 m of sediment is an awful lot of sediment to find in the deep sea so far from a continent.

In some places, the sediments are only 350 m (1,000 ft.) thick and consist of marl (lime-mud) and/or nano-fossils (skeletons of tiny marine or freshwater creatures), and a number of sapropel layers (organic-rich layers), according to Gennesseaux et al. (1998, p. 279). A study by Turkish workers on piston cores from north to south through the Aegean revealed a total of 5 sapropel layers, though all layers are not present in all the cores (Bursin et al., 2016). Thus, while we have coarse clastic sediments to the north, in the south we have fine lime-muds and organic material.

Bursin et al. mention that several such sapropel layers have been recognized across the entire Mediterranean Sea. These authors declare them to be extraordinary because: "under normal conditions, a large proportion of the organic matter in the oceans is readily oxidized (to CO_2) and consumed by bacterial grazing, so does not accumulate on the sea floor" (Bursin, 2016, p. 103). They note that previous conditions must have been significantly different with respect to water movements (hydrographic regime), surface and bottom waters, and biogeochemical cycling linked to global and regional climatic variations.

This, in my view, is throwing everything except perhaps the kitchen sink at the problem. They do tell us, though, that the bio-geochemical signature of the sapropel layers implies a mixture of terrigenous and marine organic matter, and it's the terrigenous that's interesting, since we must wonder how it got to where it is now, especially considering that it appears to be present over the entire Mediterranean.

Sapropels are thought to be due to what are called "anoxic events," which represent times when the ocean water becomes severely depleted in oxygen for some as yet, and as usual, unknown reason. The lack of oxygen means that there are no animals or bacteria available to digest the organic material, and it also prevents it from naturally decomposing due to normal oxidation. Stagnant bottom water that is not disturbed by the admixing of oxygen-rich surface water and supplied with organic material settling down will quickly become anoxic.

The bottom of the Black Sea today is one such well-known anoxic environment, and others exist elsewhere (Lake Tanganyika in Africa or Mariager fjiord in Denmark, as examples). A big part of the anoxic puzzle is the fact that such anoxic events have apparently occurred on a global scale, and it is a mystery as to how the oceans could become depleted in oxygen on such a global level.

This is yet another one of those geological puzzles we will address later. For now, though, I'll just state that if ground-up or otherwise finely divided organic material, such as macerated plants and trees, were added quickly, in large quantity, to the ocean, then buried quickly on the sea floor, it would be sealed off from oxygen and bacteria and would then, inevitably, become a sapropel layer, without any anoxia occurring, or any necessity for invoking it.

With regard to the Sapropel layers found over the general Mediterranean sea floor is an article by E. J. Rohling of the Southampton Oceanography Center, who describes this anoxic event as a "catastrophe" (by which he means it took place over a period of several centuries, like any good catastrophe described by any good uniformitarian). What is interesting is that he dates it to 9,500 years ago, which, considering the unreliability of dating systems, which we will examine later, is close enough to 11,500 years ago to suggest it occurred with the rest of the basin-filling.

In the meantime, and briefly, like everything else in geology, radiometric ages are confidently presented as downright scientific certainty. The public, of course, is not told that these ages are based on a disturbingly large number of mere *assumptions* that have not been validated, because they simply cannot be. The list of assumptions varies from four to eight depending on the writer and the particular type of radiometric dating method under discussion, but a list of four to eight is a lot of assumption.

This has led many, including myself, to derisively refer to this whole (pseudo-scientific) process as "The Dating Game." Suffice it to say for now that all these dates and long eons were pretty much settled on, in a general way, by Lyell and his uniformitarian colleagues long before radiometric dating ever arrived.

As I mentioned in the introduction, the various ages are based on the rather dubious claims and guesstimates of Lyell and his fellows. As I also said, it was this particular episode in the history of geology, and the highly unlikely coincidence of the "genius" getting it so right so early in that history, that made me initially suspicious of the whole story. However, the long ages found by radiometric dating, of course, *always* support the uniformitarian position, with its great age of the earth, and so are defended tooth and nail by the establishment.

They are, in fact, only a poor excuse for a scientific support that has been mustered to that end. Even the much-ballyhooed carbon dating, widely used in archaeology, has serious problems, as we'll see, though, perhaps, not quite so many as the others, at least back to five or six thousand years ago.

In summary, then, the Aegean has a floor topography composed of multiple basins of widely varying depth and extent, all filled with inordinately thick accumulations of recent and unconsolidated clastic sediments. Since the Aegean Sea only came into being recently—at or after the end of the Ice Age, apparently—it is difficult to account for these sediments, especially as to their thicknesses of thousands of meters and how they got where they now lie. Mind you, it's also very difficult to account for the extreme subsidence of the basins, at up to 5,000 m, which subsidence we know was also very recent.

Most workers, including the ones referred to in the foregoing, seem to simply assume or imply that the sediments accumulated over millions of years, while the former Aegean landmass, including the basins, was in the process of slowly stretching and sinking, as per orthodoxy. However, we know it wasn't stretching and sinking over long eons; Aegeis was dry land up to 11- or 12,000 years ago, and then it wasn't.

However, as we saw, those sediments were obviously transported and deposited by moving water, and a very limited number of discrete sedimentary units, or "sequences," have been identified in those basins. Typically, as we saw, various workers mention only three or four "units" and a topmost layer, representing recent or present-day deposition. What concerns me is that many, if not all, of these basins were supposedly freshwater lakes in the not-too-distant past.

Therefore, there should be some evidence of lacustrine sediments, i.e., muds and marls, in those now-submerged basins, either on top of, most logically, or intercalated with the identified clastic terrestrial sedimentary units, which are definitely not lacustrine. Most writers do not, however, mention them, and make no real effort to properly account for the presence of all the coarse, clastic material in the basins. Theoretical turbidity currents are a rather poor substitute for an explanation.

Even if turbidity currents were a valid concept, grim reality shows us that rivers do not move coarse sand and gravel even as far as the sea. In fact, there are very few rivers that have the power, even in high flood stage, to move such material even as far as their mouths. This is due to the typical slow speed of rivers as we see them in their final stretches near the coast, along with the resistance of the ocean, and, very importantly, the high tides twice a day that can back up many rivers to a standstill, and even cause backward flow.

This causes the river to drop anything coarse it's carrying well back from the coast in such cases, and is why typical river deltas are composed of silt and mud, with only a little fine sand and typically no gravel whatever. The subsequent low tide does allow the river to move faster than its average speed, but the river cannot compensate by picking back up all the dropped coarse material.

This is an example of the basic physics we're all familiar with. The exceptions to this general case are rivers that flow directly from mountains into the sea and bring coarse material with them because no flat valley or open floodplain intervenes, e.g., the Andes, Maritime Alps, among others. I'm only referring here to seasonal floods, the only time anything coarse is transported; in normal times, nothing, or just mud or organic material, is brought down by any river, even from mountains.

The general assumptions held by conventional geologists who have studied the Aegean basins is that rivers brought the material from the land and deposited it at the edges of the lakes filling the basins, where it built up over time. When it reached an "unstable" condition, an earthquake such as are common in the region disturbed the "sediment wedge" and caused it to collapse. Like any good turbidity current, it raced off down the basin slope, through the mass of static water already in the basin, and then spread itself out all over the basin floor, a surface that stretches, in many cases, over a distance of 50 miles.

As described above, no rivers bring coarse material out past their mouths and into a mass of essentially static water such as the sea or a lake. The idea that a "turbidity current" could, even if it existed, move that coarse material across the almost flat surface of a basin floor and through dead water, as far as I can see, defies the laws of physics, logic, and plain common sense.

What we find when we look at the delta or estuary of any river is a mass of material composed of fine sand, silt, and mud. We see, as any harbormaster will inform us, that this material builds up where the river meets the sea, i.e., on the riverbed or in the estuary. This is why harbors must be regularly dredged. Only the exceedingly fine material makes it further out, and this is not what we find in those Aegean basins.

If the narrative of the Aegean, as outlined above, were the correct one, then we should find only layers of fine mud and lime-mud along with marine micro-fossils as the sediments of these basins, such as those found by Gennesseaux et al. (1998) above, in the Cretan Trough, as well as dust from the Sahara and volcanic ash dispersed within it. We should find river deposits at the basin edges only, just like we find at the ocean in estuaries or deltas.

Wave action, not very powerful in lakes due to the limited fetch of the waves, may bring sand a short distance out from shore, a short distance here being measured in meters or yards, rather than miles. On any open ocean shore, like the Atlantic, for example, waves don't move sand out more than a mile or so, except during a major storm or hurricane.

Those "turbidite" deposits of various writers represent significantly more energetic events than those implied by any uniformitarian geologist, and the similarity of turbidites to flood deposits would suggest that that is just what they are, flood sediments fining upward from a coarse clastic base to fine mud on top, in a classic declining-flood profile that no one usually has any problem accepting, since it is logical, complies with the laws of physics, and especially, agrees with our own personal observations.

However, despite the complexity, the general agreement among today's geologists does, at the same time, tend to agree with that old Egyptian priest. The sole difference is that, for the priest, the subsidence, or "falling away," of the land ("soil") occurred quickly, recently, and catastrophically, leaving only a wasted remnant of the original country, or landmass, "the skeleton." I have here used my own interpretation of the word "soil" as compared to this particular translator.

This, of course, is just the same age-old catastrophism versus uniformitarianism argument that envelopes any myth or legend with a geological component involving catastrophic action, and the bones of contention are always the same: scale, speed, and date. The main point is that the same catastrophe that struck Atlantis in the middle of the Atlantic also affected Greece in the eastern Mediterranean.

It is just that it simply affected them a bit differently; Atlantis subsided beneath the waves during violent earthquakes, while what is now the Greece/Aegean region (at least) experienced some, but much lesser, subsidence, in a region known today for earthquakes and volcanoes. The implication is that this catastrophe involved rapid, large-scale tectonic activity, which, if so rapid and large-scale, would doubtlessly have caused rapid ocean water movements on a massive scale, and the ocean would inevitably have invaded the land. In turn, this would have caused widespread surface destruction, erosion, and material transport, thereby explaining the coarse clastic deposits in those basins.

In opposition to the priest's narrative of events in the Aegean, however, what we have from conventional geology is theory upon theory and no real knowledge we can claim as certain. Trying to use plate tectonics to explain the structural history of the region is fraught with problems, in the light of what the theory itself says: slow-moving "convection currents" in the supposedly ductile mantle carry the overlying rigid plates and cause them to ever-so-slowly "collide" with one another, such currents (presumably) always moving in the same direction.

The structural history of the Aegean, however, as we have seen, does not comply with that model; we see clear evidence of back-and-forth and rotational motion, and this cannot be easily explained by plate tectonics theory. This theory, as we'll see, is entirely based on unobserved and undetected processes and agents, the existence of such processes and agents being, therefore, somewhat questionable. The up-and-down motion of the region is explained by recourse to the concept of "isostasy," another entirely theoretical process. And, as we have seen from our brief discussion of seismic waves, the mantle must be solid. In which case, how could there be "mantle convection currents" in it?

Apart from the tectonic activity exhibited by the Aegean, and the volcanic activity and earthquakes that continue at present, implying that that tectonic activity was recent, the issue in the Aegean is obviously all the coarse, clastic sediments found all over it. Such sediments have been found in practically all, if not all, of the basins, the latter seeming most likely, since it would be difficult to explain the presence of a few orphans.

One must, of course, wonder how it all got so far out from land, its only source, and into those basins in such enormous quantities. We saw thicknesses of the order of two and three thousand meters and more in any basin; these sediment layers are kilometers thick, tens and hundreds of kilometers from the nearest land.

Further, the surface of the submerged central Aegean area, i.e., the Cyclades Plateau, is generally levelish, though now more uneven with islands and basins, etc. However, when it was above sea level, it was also generally level with lakes and hills and shallow valleys among them, like a sort of rolling countryside, according to many writers.

This does not sound like the kind of topography that generates large and long-sustained floods, the kind that would be at least minimally needed to get *any* coarse sediment, from *any* source, into those lakes, and, even if it reached them, the coarse material would be deposited at the edge, as is normal. There is no possibility that an ordinary river, even if large, could spread clastic sediment over the bed of any lake, and this is a problem we're going to meet throughout Greece.

We also examined another problem, that of the sapropel layers, representing deposition of organic material over the Aegean area, and apparently widely. Given the terrestrial nature of at least some of it, it can only have come from land, and, in fact, we are going to meet a lot of organic layers on Greece itself also.

The problems, of course, are resolved by positing a major flood sweeping over Greece from the northwest, stripping it of all its loose material, tearing off and grinding up any amount of rock and debris, the debris itself doing the grinding and eroding, and then sweeping all that material out over the Aegean area and filling up the basins with it. And, as we saw, to the south were found finer sediments in lower quantities, perfectly in keeping with a major flood from the northwest.

Since the typical basin contained three layers of such sediments, the suggestion is automatically made that this process repeated itself three times. We will now move to mainland Greece and see what evidence awaits us there, and what it can tell us.

Chapter References

Anastasakis, G., et al., 2006, Upper Cenozoic stratigraphy and paleogeographic evolution of Myrtoon and adjacent basins, Aegean Sea, Greece, *Marine and Petroleum Geology* 23, pp 353-369 (pp 363, 365).

Biju-Duval, J. et al., 1974, Geology of the Mediterranean Sea Basins, in: *The Geology of Continental Margins*, eds.: Burk, C. A. & Drake, C.L., Springer-Verlag, New York, Inc.

Bursin, E., 2016, Late Quaternary chronostratigraphy of the Aegean Sea sediments: Special reference to the ages of the sapropels S1—S5, *Turkish J. Earth Sci*, Vol. 25.

——— et al., 2016, Geochemistry of the Aegean Sea sediments: Implications for surface- and bottom-water conditions during sapropel deposition since MIS 5, *Turkish J. Earth Sci.*, Vol. 25, 103-105, p. 103.

Davis, W. M., 1899, The Geographical Cycle, *The Geographical Journal*, Vol. 14, No. 5, pp 481–504.

———, 1909, *Geographical Essays*, editor: Douglas W. Johnson, Boston.

Gennesseaux, M. et al., 1998, Thickness of Plio-Quaternary sediments (IBCM-PQ), *Bulletin of Geophysics and Oceanography,* Vol. 39, No. 4, pp 243-284.

Gaki-Papanastassiou, K. et al., 2010a, Palaeogeographic Evolution of the Cyclades Islands (Greece) during the Holocene, in: *Coastal and Marine Geospatial Technologies*, Chap. 28, (pp 297–304) p. 300.

Louvari, M. et al., 2019, Impact of latest-glacial to Holocene sea level oscillations on central Aegean shelf ecosystems: A benthic foraminiferal palaeoenvironmental assessment of South Evoikos Gulf, Greece, *Journal of Marine Ecosystems*, Vol 199.

Lykousis, V. et al., 1995, Quaternary sediment history and neotectonic evolution of the eastern part of the Central Aegean Sea, Greece, *Marine Geology*, Vol. 128, Iss. 1–2, pp 59–71, Abstract.

Nomikou, P. et al., 2018, Expanding extension, subsidence and lateral segmentation within the Santorini-Amorgos basins during Quaternary: Implications for the 1956 Amorgos events, central-south Aegean Sea, Greece, *Tectonophysics*, Vol. 722, pp 138-153, Abstract.

Phillips, J. D., 2002, Erosion, isostatic response, and the missing peneplains, *Geomorphology*, Vol. 45, Iss. 3-4, pp 225-241.

Poulos, Serafim E., 2009, Origin and distribution of the terrigenous component of the unconsolidated surface sediments of the Aegean floor: A synthesis, *Continental Shelf Research*, Vol. 29, 2045–2060 (pp 2045, 2057).

Riedl, H., 2004, New contributions to the geomorphology of the Aegean archipelago in particular consideration of dating peneplains and pediments and their paleoclimatic conditions, *Ann. Geol. Pays. Hellen.*, Vol. XXXX, pp 11-49, cited in Gaki-Papanastassiou, p. 300.

Rohling, E. J., 2001, The dark secret of the Mediterranean – a case history in past environmental reconstruction, http://www.soes.soton.ac.uk, visited 1-23-21.

Tsampouraki-Kraounaki, K. et al., 2012, The Santorini-Amorgos Shear Zone: Evidence for Dextral Transtension in the South Aegean Back-Arc Region, Greece, *Geosciences*, Vol. 11, p. 216.

Suess, Eduard, 1904, *The Face of the Earth, (Antlitz der Erde).*

CHAPTER THREE:

MAINLAND GREECE

THE OLD EGYPTIAN PRIEST INDICATES THAT LARGE-scale and recent tectonic subsidence accounts for the submergence of Aegeis, and this explanation, which is, in a general way, in agreement with academic geology (apart from time and speed) would account for the land of Greece not being as extensive as it once was, as the priest puts it.

However, this is something rather different from what the priest was saying with regard to the rich soil that covered the country, which was all washed away, apparently "in one dreadful night of rain or flood" and also by "earthquakes and the third terrible deluge before that of Deucalion."

So, clearly, the priest is saying that all the soil was washed away by a terrible flood that occurred over the course of a single day, just like the destruction of Atlantis itself, but he also tells us that there were three more floods since then. In which case, if true, it's no surprise that many areas of Greece today are pretty much bare of soil or only have a thin covering, as shown in figs. 3-2 and 3-3 from southern Greece and the Peloponnese, while the picture improves somewhat on moving north.

Fig. 3-1: Greek landscape, which looks generally bare of soil, which seems to only be starting to develop or redevelop in this area. Image: ID 66199157 © John Critchley, Dreamstime.com. Royalty-free license.

Fig. 3-2: View of part of a rocky hillside with much bare rock, minimal soil, and scrubby bushes. Image: ID 200052284 © MrFly, Dreamstime.com. Royalty-free license.

One can see from the figs., or from any geography or tourism book on Greece, that much of it is rather barren, especially toward the south, scrubby on the hillsides with extensive areas of bare ground or rock, as the various images show:

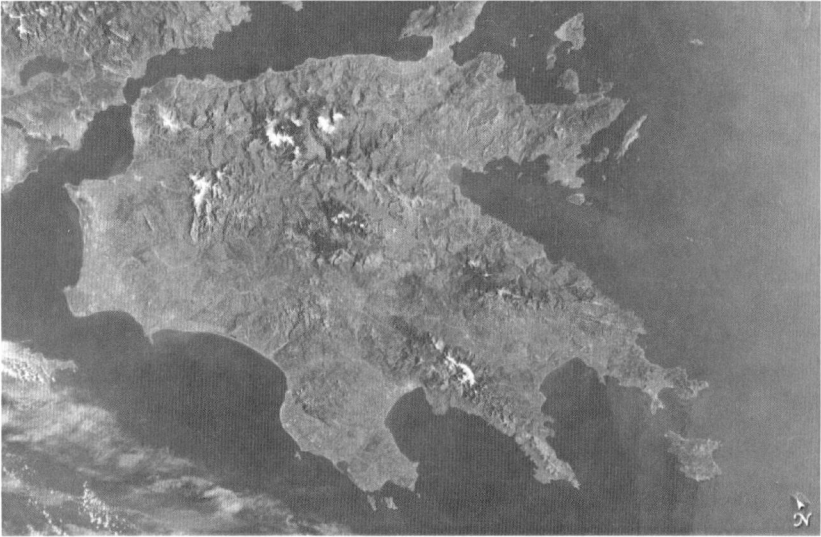

Fig. 3-3: Satellite view of the Peloponnese. The rugged terrain can easily be seen. Image: NASA Visible Earth.

Now, the fact that many areas of Greece are bare of soil today does not necessarily mean that a massive flood washed it all away. Human activities like agriculture, deforestation, and water diversion, which we know were carried on fairly extensively throughout ancient Greece, are quite possibly responsible for much of it—which is to say responsible for any soil erosion during the last five thousand years or so, since farming really commenced in the region, on any kind of extensive scale. No doubt farming went on before that, but there's not much evidence for the presence of a large population in Greece before 5,000 years ago.

A number of studies have been done on this question, given the extensive erosion, and the necessity to explain it away by some pedestrian process, which, in this case, is accomplished via the simple expedient of blaming it on us humans.

These studies suggest that it wasn't a simple, single, continuous process. E. Zangger, the geo-archaeologist we met in volume 1, did a lot of work on this question, before, that is, he got carried away with strange notions of Atlantis and wrote a book on the subject, instead of keeping his head down and his powder dry, like a sensible academic.

In his study published in 1992, he concludes that soil erosion due to early human interference mainly began about 5,000 years ago (early Bronze Age) due to farmers moving up-slope from the low plains they had hitherto focused on, probably because of increased population. He further concludes that it occurred in different places at different times, subsequent to this. (Zangger, 1992, *abstract*, p. 133). So, Zangger thinks the period around 5,000 years ago is significant in terms of farming that led to erosion, meaning something noteworthy must have occurred at that time, just like many other places we've been.

This first phase of erosion was most significant in terms of the quantity of soil removed (Zangger, 1992, p. 135) leaving the hills bare and thick layers of alluvium (flood-deposits) on the plains. These layers of alluvium overlie and taper out against older Pleistocene deposits that form the base fill of the valley floors and plains, see map in Zangger (1992). These two materials, by the way, sound much like Bald's new and old alluvium and Buckland's alluvium and diluvium. Needless to say, Zangger assumes that all this soil erosion took place over time, as the hills were cleared of trees.

It must be said, however, that when viewing the areal extent of the bare hills and barren rocky landscapes of southern Greece, those ancient Greek farmers were most exceedingly and surprisingly thorough in their deforestation efforts, over what Zangger concludes was a rather brief period of a few hundred years, from the middle to the end of the 3rd millennium BCE, or 4,500 to 4,000 years ago.

He also notes a distinct (sudden) change in climate (Zangger, 1992) with oak forests being replaced by hornbeam, pine, and a shrub called *erica* (tree heather). The olive tree is also found, but it is not known whether it was wild or domesticated. A sudden change in climate certainly sounds like something noteworthy occurred.

Zangger does not think deforestation alone was responsible for all the erosion, but that expanding pastoralism prevented plant recovery. While deforestation, pastoralism, and, of course, cultivation will result in soil erosion, and I would agree with Zangger as far as the lower hill slopes are concerned, the steeper slopes and the mountain tops are a different matter though. Such slopes are way too steep for any but mountain goats and we can safely assume that no farming went on the hilltops, no more than it could today.

Trees and scrubby bushes are obviously recolonizing the mountain slopes in fig. 3-4. The tree-line can easily be discerned below the more or less bare rocky hilltops in the left distance, though we can see lower down that patchy grass is developing with it. Given that the mountains are obviously being reclothed with trees and grass, the question arises as to how long it's been going on, and why all the mountains we can see here, near and far, are all in a similar state of undress. Considering Greece is mostly mountains, and there's a lot of them, it really seems a bit much to blame a few Stone Age farmers for stripping multiple mountain ranges in Zangger's few hundred years. To my mind, that kind of work needs a lot more people or much more time.

Fig. 3-4: Peloponnese, Greece. These mountains are a bit high, bare, and steep for farming to be very feasible. Image: ID 119723614 © Bruce Whittingham, Dreamstime.com. Royalty-free Lic. (Cropped).

As well as that, however, it is well known, and long-studied, that when rain, for instance, washes away soil from a hill under cultivation, that soil is usually just washed down to the base of the slope, and oftentimes it's just banked up against obstructions on the slope itself, such as stone walls or some natural feature. In any case, there generally being few or no rivers skirting around the slopes of most hills, the soil should just be heaped up at the base of the slope.

That is not what we see here in Greece, and no obvious soil heaps or wedges were reported by Zangger or anyone else. What soil was on those hills has been removed somehow, far from the slopes it used to cover, and certainly not by any everyday rain shower. Apart from that, anything washed down is usually incorporated into soil at the slope base and naturally generates plant growth.

We know that soil tends to develop its way *up* a mountain and not from the top down. Weathered out and eroded particles, as well as wind-borne dust, etc., get washed down and become trapped by plants on the lower slopes. They then get incorporated into the soil, pushing the soil-front upslope, whereupon new plants colonize the new soil, and thus soil, plants and trees advance uphill. Since all the mountains are similar and the soil and plant growth does not seem to have been going on very long, I have to conclude that these mountains were washed bare by something in the not-too-distant past and are still recovering (pun not intended).

Fig. 3-5: Pindus Mountains, northern Greece (2,637 m, 8,652 ft.) The slopes and lower hilltops are well-covered with plant- and tree-growth, unlike the Peloponnese mountains shown above. Image: ID 31249356 © Lefteris Papaulakis, Dreamstime.com. Royalty-free Lic. (Cropped).

The high mountains of northern Greece, on the other hand, exhibit much better developed soils and much more luxuriant plant growth on their flanks, though they also tend to have bare hilltops. Figure 3-5 is a view of the Pindus Mountains (peak height, 8,752 ft.,

2,660 m) showing good plant, shrub, and tree growth up to a considerable height, which can be guesstimated at about 5,000 or 6,000 ft. (1,500 or 1,800 m) or so, based on the height of the snow-capped higher peaks in the distance.

Northern Greece has a more continental climate with cold winters, snow, frost, and rain and with a generally wetter climate. Thus, the northern mountains are subject to much more weathering and erosion than those of the Peloponnese. In such case, they are more efficient generators of the debris and fine material that would go into soil formation. Rainy conditions would tend to prevent light material from being blown away, while fostering plant growth at the same time. And, as we can see, the slopes shown in fig. 3-5, though steep, have better developed soils and more plant growth, and hence would be expected to be more suitable for farming and grazing than their southern counterparts, at least for sheep and goats.

At the same time, however, given the limited population of the time, it is not at all likely they needed to farm the mountainsides 5,000 years ago. And they certainly wouldn't need to farm the mountaintops, and one does have to wonder how much wood the few people would have needed in a relatively warm climate.

Zangger was also a coauthor of a paper with T. van Andel and A. Demitrack on the same subject (published in 1990, dated as 1991), and both papers agree on an early and major erosional "event" that laid down flood deposits (alluvium) on the older Pleistocene deposits of Ice Age times. These younger deposits consist of coarse, poorly sorted sands and gravels, similar to the underlying Pleistocene, which were supposedly deposited during the Ice Age. This "alluvium" I must say is considerably coarser than the usual fine sands, silts, and clays we normally consider alluvium to consist of, and these deposits sound a good deal more like "diluvium."

The former comprised the old post-glacial land surface of Greece and was capped by a dark-brown soil that shows evidence (roots, etc.) of apparently rich plant growth. This soil was later eroded away by some widely acting agent or process, except for places where it was buried under alluvium. While they don't say so explicitly, it is understood that the agent that eroded away the soil was the same agent that deposited the overlying gravelly "alluvium."

According to van Andel et al. (1991), and close to Zangger's (1992) age-date, this "alluviation event," as van Andel terms it, occurred sometime in the late 4th millennium BCE, and most likely close to 3000 BCE (5,000 years ago). According to dates obtained from samples above and below this material, it was a short-term event lasting no more than a few centuries (van Andel et al., 1991, p. 382).

Since van Andel et al. only supply dates from above and below this layer and not from within the layer itself, this event may have been much shorter-lived than a few hundred years, especially since this alluvium included cobbles, pebbles, gravel, debris-flows and stream-flood deposits, all indicative of lots of water moving quickly. Further, they state in their discussion that this early phase (event) of erosion was greater in areal extent and landscape effect than all subsequent phases, or episodes, of alluviation (van Andel et al., 1991). Importantly, however, all of the foregoing, sounding all kinds of reasonable, belies some kind of clearly extraordinary event having occurred about 5,000 years ago.

While the foregoing deals with the more recent erosional and depositional history of Greece, and with soil development and human interference, etc., it would be wrong to assume that only ongoing processes are necessarily responsible for the Greece we find at present, and van Andel et al. acknowledge that latter-day alluviations may have had little to do with the landscape we see today.

They instead suggest that most of the landscape development was accomplished during the Quaternary/Holocene changeover to non-glacial conditions, with all its dramatic climate changes, tectonic activity, and so on. In other words, they are referring to the end of the Ice Age, which, as we know, was the time that Atlantis disappeared, and the priest said Greece was also overwhelmed by a flood. And these writers are here acknowledging that the end of the Ice Age involved some "dramatic events," which do not sound at all uniformitarian.

The authors refer, toward the end of their paper, to work done by Hutchinson (1969) and Rohdenburg & Sabelberg (1973), who claim that it was during this period that the Greek landscape was given its present form, and that nature had a lot more to do with it than us humans.

From an objective and more practical perspective, this seems fairly reasonable, since serious human interference must be a much more recent phenomenon, given the small populations that supposedly existed in early Greece, and the limited influence those people could have had on the landscape, no matter how wantonly destructive they were.

Thus, taking all the foregoing together, it would appear that the Greece/Aegean region was subject to a period of considerable erosion and deposition about 11,000 to 12,000 years ago, and it suffered another major period of erosion and deposition, an "alluviation event," at about 5,000 years ago, according to uniformitarian geologists.

Regardless of recent events and sediments in the Greek/Aegean area, what we are most interested in is what occurred on Greece at that period referred to in the last paragraph, the end of the Ice Age, and apart from anything else, this period left its mark on the country in the form of sediments that underlie the distinctly recent, or Holocene Epoch. These sediments are dated as Pleistocene in age, supposedly having been deposited by glacial ice over the usual long ages of hundreds of thousands or even millions of years, and Greece is considered to have had a number of glaciers in the high mountains.

While I would easily allow glaciers in the high mountains, given the present snowfall, the areal extent of Pleistocene sediments is, as we'll see, much greater than the rather limited extent that glaciation could possibly have had, given that Greece is really too far south for much Ice Age effect. Thus, fluvial action from meltwater streams, etc., has been invoked to account for those sediments that are not directly ascribable to glacier action, even though they look more or less the same.

As usual, the question will boil down to some local version of the usual one: was it slow, incremental deposition by creeping glaciers and low-volume, Ice Age meltwater streams, or was it the rapid deposition of sediments by large volumes of water moving quickly, as in the kind of "alluviation event" referred to by the authors above? In other words: low-energy, long-term or high-energy, short-term?

We must begin with a general overview of the Pleistocene Period, or Ice Age, as it pertains to Greece, focusing on the sediments and landforms, the geomorphology, as it's called. The Greek landscape consists of three general types, the mountains, the valleys, and the plains, as well, of course, as the coastlines.

Fundamentally, the science of geomorphology is all about what processes and agents shaped and molded any landscape to its present condition, and there aren't a whole lot of them; rain, rivers, wind, frost, and ice are more or less the important ones. In Greece, as in much of the northern hemisphere, or just the northern half of it, the main geomorphological agent is considered to have been large volumes of moving ice, rather than wind, rain, and rivers, though they have supposedly played a major part.

Obviously, large volumes of moving water are never considered, since floods are not, of course, allowed. All the ice is in the form of ice sheets, ice caps, ice fields, or just individual glaciers, and Greece suffered like everywhere else. In other words, despite its excessively southern latitude, compared to Britain for example, Ice Age Greece would appear to have somehow suffered downright polar-like conditions, at least according to the (mostly British) glaciologists who studied it, as we'll see in the following chapter.

While I consider such amounts of ice on Greece to be impossible, Ice Age or no Ice Age, given the latitude, it occurs to me that a good comparison in the present day would be with southern Norway.

It is near the arctic circle, gets plenty of moisture, but has rather limited glaciation in the form of mountain glaciers, despite its high northerly latitude. Hence, the present state of the world would suggest that Greece had little enough ice on it during the past Ice Age, it being well-south of the estimated ice limit. I, personally, would be inclined to only allow a few glaciers on the highest mountains, and only in the north, and, realistically, little more than that.

Since the Ice Age ended almost 12,000 years ago, we must wonder at the rather bare and rocky surface of Greece. While there is no doubt that it is somewhat arid, and it is possible that wind and rain have removed a good deal of material, there still remains a question as to the poor soil development in that 12,000 years. The priest is very explicit when he says that Greece (Athens) was blessed with lots of good, rich soil, even on the hills, with abundant tree growth, luxuriant plant life, etc. This description obviates the possibility that rain had anything to do with the removal of the soil, since rain, no matter how heavy, has no effect whatever on soil covered with plant life.

Despite what geologists say about rain's power to erode, we all know from our own experience that the heaviest rainstorm will not wash away anything from a hill covered in grass, for example, and all we ever see in the rainwater that runs off is some fine mud or clay, that may have been, as always, exposed to its effects in some minor grassless places. We must also wonder at rain's claimed ability to erode hard rocks, and everything else of which it's supposedly capable.

Hence, that only leaves two totally opposed possibilities, massive and overwhelming flooding, or the Ice Age, that great agent of destruction that rendered the idea of deluges, such as that of Noah or Deucalion, obsolete and unnecessary. At the same time, a deluge would solve a lot of problems, both on Greece and in the Aegean. We will, therefore, begin with the Ice Age on Greece and first see if it was up to the job.

Chapter References:

Hutchinson, J., 1969, Erosion and Land Use: The Influence of Land Use in the Epirus Region of Greece, *Agricultural History Review*, 17:85-90, Cited in van Andel et al. (1991).

Rohdenburg, H. & Sabelberg, U., 1973, Quartäre Klimazyklen im westlichen Mediterrangebiet und ihre Auswirkungen auf die Relief- und die Boden-Entwicklung, *Catena*, 17–180, cited in van Andel et al. (1991).

Van Andel, T. H., Zangger, E. & Demitrack, A., 1991, Land Use and Soil Erosion in Prehistoric and Historical Greece, *Journal of Field Archaeology*, Vol. 17, No. 4, pp 379-396 (pp 384, 389-390).

Zangger, E., 1992, Neolithic to Present Soil Erosion in Greece, in: *Past and Present Soil Erosion, Archaeological and Geographical Perspectives*, Eds.: Bell, M. & Boardman, J., Oxbow Books, Oxford, The David Brown Book Company, Bloomington, Indiana, U.S.A.

CHAPTER FOUR:

THE ICE AGE IN GREECE

THOSE WHO HAVE STUDIED THE ICE AGE IN GREECE concur that glaciation was limited to the highest parts of the mountains, mostly above 2,200 m (7,200 ft.) but possibly down to 1,800 m (6,000 ft.) (Leontaritis et al., 2020). These authors tell us that Greece did not boast much in the way of glaciers or ice caps. Greece is, after all, very far south for an Ice Age to affect it, but, given that the Ice Age itself is poorly understood, especially its more southern and secondary effects, ice on 7,000 ft. (2,130 m) high mountaintops can easily be allowed, especially on the Pindus mountains (8,700 ft., 2,650 m) and on Mt. Olympus (9,500 ft., 2,900 m), both in the north.

The actual northern areas of the Pindus mountains are considered to have the most extensive glacial landforms, particularly the higher parts of Mt. Grammos, Mt. Smolikas, especially Mt. Tymphi, and Mt. Olympus to the east.

To the south, in the Sterea Hellas, traces of extensive glacial valleys and (glacial) till, as well as high-altitude moraines within numerous cirques, were reported on Mt. Parnassus, and Mt. Agrafa also shows traces of former glaciation (Leontaritis et al., 2020, pp 77-78). On the Peloponnese, the focus is on Mt. Chelmos, which, apparently, is the only one that has been seriously studied. The authors report, from earlier studies, the presence of cirques, moraines, perched boulders and ice-molded bedrock, which are features that are generally attributed to (Quaternary) glaciation, extending down to 1,850 m (6,000 ft.).

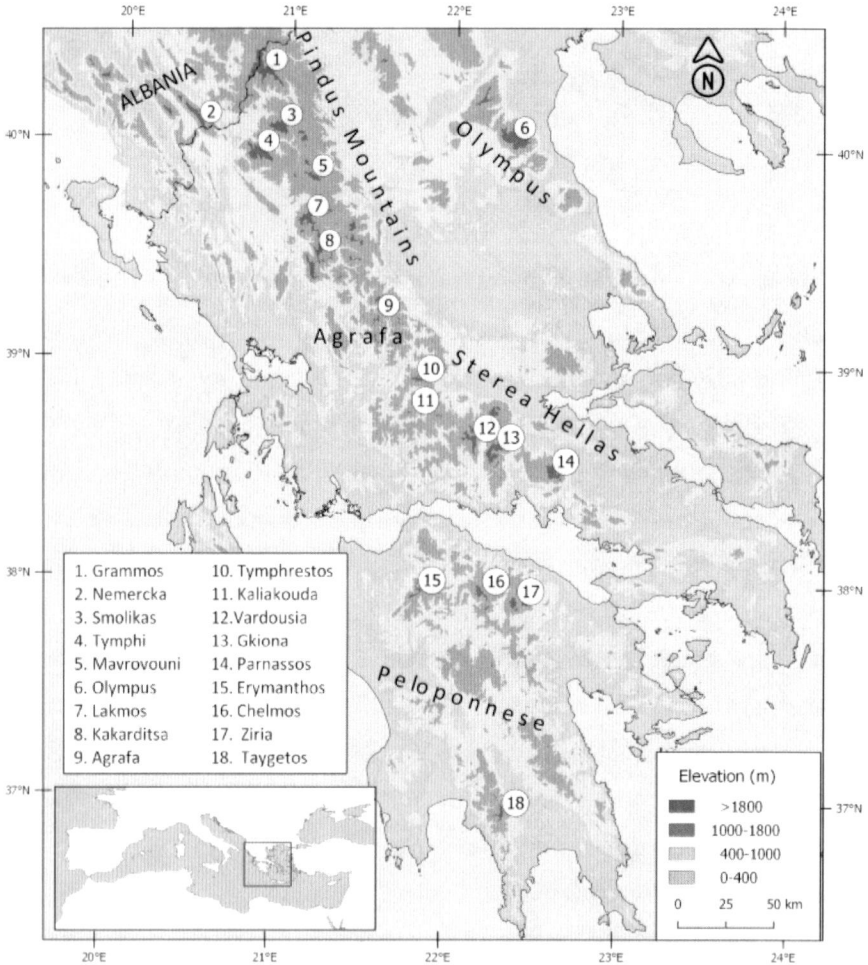

Fig. 4-1: Mountains of Greece: Numbered mountains are those with supposedly identified glacial evidence. I would be extremely skeptical of glaciers on the Peloponnese, so far south, or even in the Sterea Hellas. Image from Leontaritis et al. (2020) p. 66: from various sources listed.

However, deposits interpreted as tills have been found down to 1,200 m (3,900 ft.) and with alluvial fans further down, which raise a few questions. Traces of glaciation have also been claimed as present on Mt. Ziria and Mt. Erymanthos and even further south on Mt. Taygetos, the latter being at the southern end of the Peloponnese (Leontaritis et al., 2020, pp 74-75), all of which glaciation I seriously doubt

While most workers deny glaciers as far south as Crete, two (Nemec & Postma, 1993) argue for their former existence based on the presence of alluvial fans (formed where floods, of any kind, debouche from a mountain valley out onto a plain) that they claim were produced by: "large water discharges associated with ice-melting in the White Mountains," the highest on Crete (Nemec & Postma, 1993, cited in P. D. Hughes et al., 2006, whom I mentioned in volume 1 regarding floods on Crete). And, clearly, Nemec & Postma are treating flood evidence as a proxy for the presence of glaciers. Overall, therefore, Greece had a very limited area of ice cover, and thus we should not expect evidence of extensive glacial erosion or deposits.

It must be remarked here that the glacial evidence mentioned in the foregoing and generally thought of as diagnostic of former glaciation/Ice Age has, in every case, an alternative explanation, and the origin of till, cirques, moraines, and erratic boulders has been, and still is, to some extent questioned. It was much more hotly debated in the early days of the Ice Age theory, however, like much else.

While most geologists now more or less accept these features and deposits to be of glacial origin, it is significant that seashells and the remains of animals and plants have been found within supposedly glacial tills in many localities, meaning that these tills clearly weren't formed under a glacier where shellfish, plants, and animals do not typically reside. Hence some tills must have a different origin.

Cirques are found in places never glaciated, as are various grooves, striations, rock sculpting, erratics, and many other features, especially erratic boulders. In his standard text, *Glacial And Quaternary Geology*, the famous (late) professor, R. F. Flint, cautions on page 141: "None of the features of a glaciated valley is in itself diagnostic of glaciation. Each individual feature is found in regions never glaciated." He then goes on to say: ". . . the combination of these features has never been reported from a valley not glaciated, and, as each feature, if not glacial, is the product of a different cause, the occurrence of all of them in a single non-glaciated valley is improbable." (Flint, 1971, p. 141)

Now, the occurrence of *all* of the typical features considered to be of glacial origin in a single valley, even if it was formerly glaciated, is a lot to ask for, and I would dare any geologist to find them. What does one say when only one or two of these features are found? Flint (1971) finishes his caution with the rather subjective suggestion that: "The weight of evidence, rather than the presence of a single diagnostic feature, justifies the inference of glaciation."

Therefore, if, in any particular valley, a plurality of glacial-seeming features is ascribed to the overarching cause of glaciation, but at the same time each type is known from elsewhere to *also* have a different cause, then it's perfectly possible that a plurality of these same features, when found in *any* particular valley, may also due be to one overarching but *non-glacial* cause, and each one is not ". . . a product of a different cause . . ." as Flint puts it. In which case there is a danger of seeing ice where no ice ever was, such as Crete, for example, and a glacial diagnosis might just be force of habit regarding poor evidence.

Hence, there is also the danger of mistaking the presence of only one or two features as indicative of former glaciation. There is, of course, also the possibility that glaciation *and* some other agent or process were at work in the same valley at the same or different times. While this is not the time or place to get into an extensive discussion of glaciation or the Ice Age, it must be remembered that all of these questions were debated since the early-middle 19th century and much indisputable evidence was presented against many of the interpretations (let alone explanatory theories) that are current among today's glacial geologists. We will fully address all these questions in later volumes.

Beginning with the Pindus mountains, early studies concentrated on Mt. Tymphi. A number of supposedly typical glacial features have been recorded: cirques, "ice-scoured" troughs, limestone bedrock pavements, rock glaciers, and extensive, well-preserved moraines, including lateral, end, and hummocky. A karstic landscape is also present, and Leontaritis et al. (2020) refer to it as a "glacio-karst" system, which I assume is a karst landscape supposedly modified by glacial action, either before or after it was karstified.

The Burren area of Ireland is considered a classic karst example, though karst features can only be produced by water dissolution, and thus, neither glaciation, nor ice of any other kind, is involved in its actual formation. Therefore, ice can only have been involved in the modification of the water-formed landscape.

The glacial features found in Greece are all high up in the mountains, but claimed traces of glaciation have been recorded much further down the valleys, as far down as 700 or 800 m (2,300-2,600 ft.) above sea level (a.s.l.), as shown in fig. 4-2. It is assumed that typical valley glaciers existed here based on what are considered to be glacial deposits such as moraines and tills, etc., in the areas around the villages of Tsepelovo, Skamnelli, and Vrysochori (see fig.

4-2), all of which lie at about an elevation of only 1,000 m (3,200 ft.). The tills are exposed in a number of road-cuttings which reveal their makeup, so we can see just how till-like these tills really are (Leontaritis et al., 2020, pp 69-70).

Fig. 4-2 is a 3D rendering of the Mt. Tymphi plateau area, showing the estimated, or conjectured, glaciated area, as well as the villages named, and the main rivers, the Aoos and the Voidomatis, which drain the area, and presumably transport away much of the material washed down from the formerly glaciated areas.

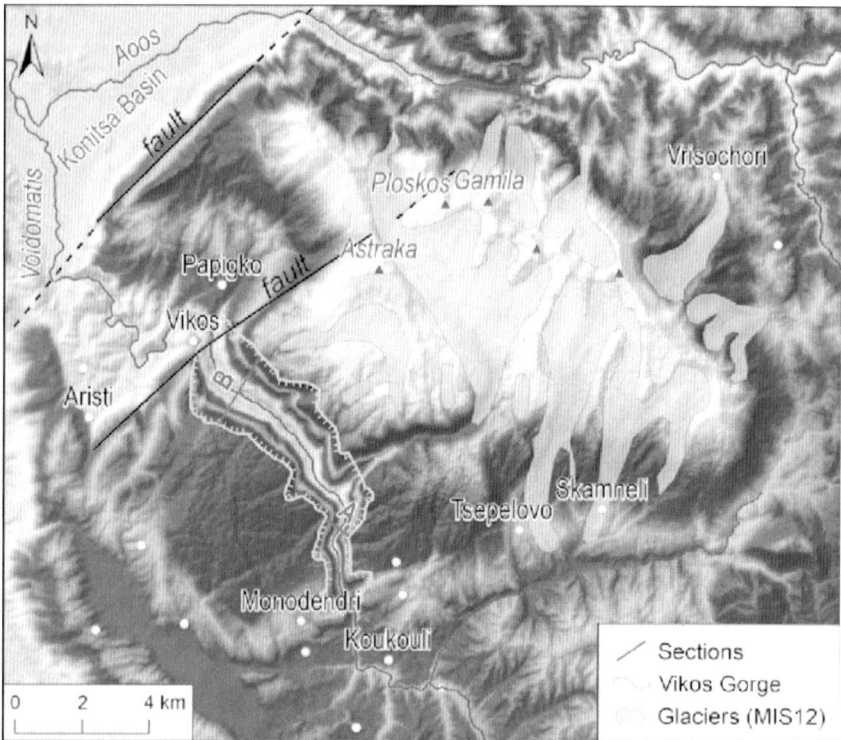

Fig. 4-2: Map of Mt. Tymphi area showing glaciers, villages, rivers, and Vikos Gorge. Image: Telbisz et al. (2019) (glacier extent taken from Hughes et al., 2007).

It is rather interesting that the few authors who have seriously studied Mt. Tymphi agree that there were three phases of glaciation; something of a coincidence considering that the Egyptian priest, and the ancient Greeks themselves, all agreed that Greece had suffered from three or four major floods, and not to mention what we saw in the Aegean earlier. Academia, needless to say, claims that all the major erosion and deposition were accomplished by glaciers and the meltwater streams flowing from them.

If so, then we should find moraines and tills left in the mountain valleys, and the same material also washed down the meltwater stream channels and out onto the plains. The material washed down, of course, should be much more rounded from transport than that left in the mountains, which should be sharp and angular, having only been fragmented. Any fine material will be carried through the alluvial plain to the sea and be deposited at the shore, as we discussed earlier.

The most interesting deposits are therefore the moraines, tills, and fluvial deposits, including whatever alluvial fans, if any, are present at the valley mouths, or elsewhere, which might be expected from hundreds of thousands of years of meltwater floods. The actual origin of the "till" or moraines is rock fragmentation due to frost action, which supposedly affected the heights above the ice. The debris produced would fall onto the surface of the ice to be transported away to melt out of the glacier as a moraine.

Shown in fig. 4-3 are three groups of different areas delimiting three glacial phases, as discerned by Hughes et al. (2007), again paralleling, with ice, the claims of the ancients, and echoing the arguments of the 19th century over whether the evidence represented a fast-moving deluge of water, or a slow-moving "deluge" of ice.

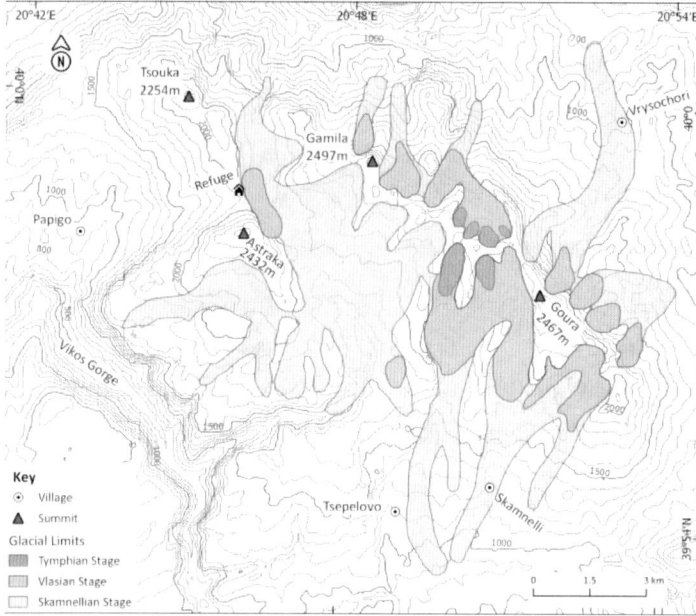

Fig. 4-3: Different glacial stages on Mt. Tymphi. Image: Leontaritis et al., 2020, p. 69. (edited from Hughes et al., 2007).

In other mountain areas of Greece, thought to have been glaciated, the same features are reported from the summit areas but are not as extensive as those on Mt. Tymphi.

On nearby Mt. Smolikas (see fig. 4-1) glacial features are less obvious, and seem to be restricted to the north side, which is rather odd in that signs of glaciation should be more or less present on each side, since temperature is elevation-related and I can't see how the necessary snow could have particularly favored the northern side.

On Mt. Olympus to the east, clear evidence of glaciation is reported, Leontaritis et al. (2020, pp 68-70). Further south, on Mt. Agrafa and in the Sterea Hellas, the same evidence is reported as elsewhere, i.e., cirques, moraines, glacial valleys, tills, etc. In the Peloponnese, Mt. Chelmos and Mt. Taygetos are both considered to have been glaciated, though there is some debate about the other mountains in the area (Leontaritis et al., 2020, pp 74-78). According to Pope et al. (2015) Mt. Chelmos was covered by a plateau ice field.

This latter would be something like the situation on Mt. Tymphi, shown above, with its valley-glaciers radiating out from the summit area. In most, if not all, cases, three glacial phases, or stages, have been identified by the various workers, while a minor, fourth one, has been claimed (P. D. Hughes et al., 2006) for Mt. Smolikas.

Applying R. F. Flint's criteria, these various workers would appear to have found at least two, or more, diagnostic pieces of evidence for the former glaciation of the highest parts of the highest Greek mountains, with cirques and various moraines being the most commonly mentioned. Striated and molded bedrock are reported on only two, i.e., Tymphi and Chelmos. Tills are reported in all cases. These, however, along with lateral and end moraines, are well down in the valleys and, as mentioned earlier for Mt. Tymphi, exposed at various elevations down to 800 m (2,600 ft.) or even lower.

Two or even three supposedly "glacial" features are really not a lot to go on, and while we can fairly easily allow glaciers to have occupied the mountain heights, according to Woodward & Hughes (2011, p. 175), there is only "very limited evidence for the transport and deposition of glacial sediments to lower elevations." They do, however, show a map of the Mt. Tymphi area with some glaciers extending down along the valleys, their evidence apparently being supposedly typical "glacial" sediments such as tills, boulders, and moraines, now supposedly lower down in those valleys (Woodward & Hughes, 2011, p. 189).

What strikes me as a little odd is that the glaciers shown in figs. 4-2 and 4-3 display a very irregular distribution, in terms of the elevations of their termini, with limits shown varying from 700 m to

1,600 m (2,300 ft. to 5,200 ft.) a.s.l. over the very small area of only 200 square km, and only about 15 or 16 km north to south.

In any glaciated area, the glacier melt line must always be at approximately the same elevation for all glaciers, or decreasing gently to the north, since the air temperature will be about the same at any particular elevation. Hence, multiple glaciers in any area should all have their termini at about the same elevation, technically known as the Equilibrium Line Altitude (ELA), which will move as a whole with any change in general climatic temperature (Ohmura et al., 1992, p. 397).

In most glaciated areas today, the elevations of the glacier termini do not vary by any significant amount over the short distance of only 15 or 16 km, as is the case above with Mt. Tymphi. In the Alps, for example, the ELA varies over a N-S distance of 100 km by only 220 m (Evans, 2006), and therefore the spread of 900 m claimed by Hughes et al. (2007) is very questionable. Since the Alps only show a 2.2 m (7 ft.) variation per kilometer, the glacier termini on Mt. Tymphi should not vary by more than about 20-25 m (60-80 ft.).

As an aside, with global warming today, glaciers all over the world are retreating, and are doing so in parallel, all retreating evenly in any given area, as is logical, if lamentable, though they are revealing much of interest regarding glacier behavior and its impact on the landscape.

The extent of the glaciers shown in fig. 4-3 above is based on identifying tills or moraines, primarily, down the valleys to the elevations indicated on the map (Woodward & Hughes, 2011, pp 181-183). Since nothing has ever been simple in the science of geology, there has been quite a debate about the various types of sediments ascribed to glacial action. They can be divided into three classes, those deposited by the glacier itself right at the end or sides (strictly glacial), those

transported and deposited by meltwater streams coming from the glacier (glacio-fluvial), and those transported by rivers, not necessarily
draining a glacier but accepting tributaries that do.

Such tributaries bring glacio-fluvial sediments with them, as
well as anything else they may pick up along the way, and these are
further reworked by the river, and deposited in the riverbed as fluvial
sediments or outside of it as alluvial sediments. Alluvial deposits are
also typically found where meltwater streams exit mountain valleys,
and dump their sediment load directly out onto the plain. Here sediments can be very coarse with boulders, gravel, sand, etc.

Till is another matter, and it is clear that not all tills are deposited by glaciers, the original interpretation. Similar deposits formed
by some other means have often been mistaken for till. It can be very
difficult to distinguish among glacial till, debris flows, landslides,
(glacial) outburst floods (jökulhlaups), or even just simple floods,
flash or otherwise, that carry a heavy bed-load or bottom slurry.

The following photos demonstrate some of the difficulties in
diagnosing the type of a given deposit and its origin. The first photo
is of a "till" from the west coast of Wales, UK, while the second photo
shows a supposed flash-flood deposit of alluvium from Bear Creek,
Colorado, U.S. The similarities are obvious, and just as obvious is
the difficulty in identifying which is which.

The general, and non-genetic, term for till and all the other
deposit types listed above, is *diamicton*, which simply means a mixture of two distinctly different material types, as in sizes, being usually
comprised of larger fragments in a fine matrix, and with little grada-
tion between them, but some variation in the size of the fragments. Its
actual composition is somewhat nonspecific in terms of any defining
parameters, and it is also generally described as poorly sorted sediment having a wide variety of grain sizes (Brevik & Reid, 2000).

Fig. 4-4: Till or Boulder Clay, Wales, UK, Eric Jones/
Waterlogged Till/CC BY-SA 2.0. (Cropped).

Fig. 4-5: Identified as flash-flood deposit.
Image source: D. A. Lindsey, USGS.

Figure 4-6 is an example of a diamicton that is very different from the two shown above, with very well-rounded cobbles, stones, and pebbles in a sandy/muddy matrix. This has been identified as a glacial till, despite the fact that the clasts are so well-rounded, a shape they can get only by being transported by water, and, considering the size of the cobbles, it must have been a lot of water, and moving quickly.

Fig. 4-6: So-called glacial till from the Snake River, Wyoming, which looks decidedly un-till-like. Image: Dr. John Ellis, academic.missouriwestern.edu.

There is no way I could accept this as a glacial till, or as having had anything to do with ice. This, to me, and without question, is a flood deposit, and some crude oblique layering and orientation of the cobbles is observable, which is most likely crude crossbedding by strong water flow from left to right (rather than imbrication, i.e., head-on-toe overlapping, by the opposite). I would guess that this deposit represents the bed-load of a major flood.

Figure 4-7 shows a till deposit or, strictly-speaking, a "diamicton" which has been identified as a glacial till, on the north side of Mt. Tymphi (Woodward & Hughes, 2011). While it is clearly a diamicton, the presence of a considerable number of sub-angular to sub-rounded boulders and stones, which outnumber the more angular ones, would tend to conflict with the claim of ice-origin and transport *only*.

Ice does not roll boulders and stones along, nor does it bang and bounce them against each other; this mechanical, attritionary process, the only way to round rocks, can only occur in a dynamic fluid such as water, or air perhaps in the case of sand or small pebbles. It is much the same process as occurs in a lapidary machine, for example. We all know how a glacier moves, i.e., at a "glacial pace," and hence there's no current or turbulence; thus, it would seem that there was more than just ice transport involved here.

Fig. 4-7: Roadside cutting in so-called till sediments showing wide range of boulder sizes in a fine limestone sand-silt-clay matrix. Image: Woodward & Hughes, 2011, p. 187.

The only thing any glacier does is transport material on its top or within itself and, at most, it grinds fragments of rock that are packed closely together within the glacier, either against each other or against the valley sides or floor, and it does this very slowly indeed. Ice on its own has no ability to abrade solid rock, including the hardest ice of central Antarctica, as it is much softer than any rock.

Ice is only a very viscous liquid/plastic solid, and so it has a great ability to transport rocks, stones, etc., but they must be concentrated and in close contact for the ice to cause them to grind against each other. The most that glacial grinding action can do is knock the sharp corners off the chunks of rock it's carrying or flatten one side by sliding it against any adjacent bedrock.

Any examination of the material deposited at the terminus of a modern glacier shows that the (moraine) fragments are mostly angular to subangular. Subrounded stones, along with any rounded ones, usually come from the meltwater streams running beneath the glacier, rolling the stones along like any river. Sand-grade material is often abundant, but there is sometimes little fine material, such as mud or silt, as it tends to be carried away by the meltwater.

As was often pointed out in the old days, and despite a superficial resemblance at first glance, close inspection reveals that the so-called glacial tills or moraines of the Ice Age show little resemblance to recent glacial deposits. Shown in fig. 4-8 is a modern moraine that obeys this rule; it looks similar to the above "till," but the stones are mostly angular with a sand/silt matrix. The same goes especially for a material known as "boulder clay," which was long thought to be synonymous with "till" and was a common term for it in the old days. It is essentially indistinguishable from typical Ice Age till, but is now, like till itself, known to also derive from other processes.

Fig. 4-8: Glacial till, Sierra Nevada Mountains, California. Image: Wikimedia Commons, Photo: James St. John, CC Attribution 2.0 Generic.

Following in fig. 4-9, on the other hand, is a photo of what Woodward & Hughes (2011) call a moraine with "limestone-rich till" sediments from the southwestern side of Mt. Tymphi, on the road between the villages of Tsepelovo and Skamnelli at an elevation of about 1,100 m (3,500 ft.). Compared to the till in fig. 4-8, it is easily seen to differ substantially. The boulders in the foreground, as well as others discernable in the face of the cutting, show a considerable degree of roundedness, as well as others which are subrounded and some that are angular.

Hughes et al. (2011) on page 182, who got a closer look at it than we get here, tell us that it is a Mt. Tymphi till with "sub-rounded limestone boulders, cobbles and gravels in a fine-grained matrix rich in sands, silts and clays." If so, what rounded off the boulders and

cobbles, especially the prominent ones clearly to be seen in the cen-
ter-left of the lower foreground? Water must have been involved at
some point, either as a flood, as lubrication for a debris flow (lahar) or
as a destabilizing agent facilitating an extensive landslide. Whatever
the case, only movement of some kind, usually transport by water,
rounds off boulders.

Fig. 4-9: Glacial sediments exposed in a quarry on the Tsepelovo
to Skamnelli road showing the rounded form of the oldest
moraines on Mt. Tymphi and the limestone-rich till sediments.
Some boulders look a good deal more rounded than Hughes's
"sub-rounded" implies. Image: Hughes et al. (2011) p. 185.

Thus, according to these writers, this is a moraine composed
of till sediments, and while they provide a list of glacially diagnostic
features from the heights of the mountain, this is the only evidence
they mention for the lower parts of the valleys. As Flint (1971) cautions, one piece of evidence is not enough, a plurality is needed,
and the weight of evidence must be on the side of glaciation. And,
by themselves, "sub-rounded" or roundy boulders are not weighty

evidence for glaciers; they're weighty evidence for the only thing that can produce them—a flow.

As mentioned earlier, most workers identify three glacial stages in Greece and, Leontaritis et al. (2020) refer to a number of papers by Hughes and his coworkers, that describe the tills as containing: ". . . sub-rounded limestone boulders of a wide size range, cobbles and gravels commonly dispersed in a fine-grained matrix rich in sands, silts and clays . . . analyses of the diamicton sequences within these moraines indicated that stacked diamicts separated by gravels related to melt out and glacial retreat record at least three phases of former glacial advance and retreat."

Thus, we have three layers of till (or diamicton or moraine; the writers make no distinction) lying one above the other, like any depositional series, with a clear break between each, obviously indicating three episodes of deposition separated by periods of quietude of unknown length, or else no separation at all but quick succession.

Fig. 4-10: Exposure of layered till (moraine) near
Tsepelovo. Image: J. C. Woodward (Twitter).

The photo of fig. 4-10 was posted to his Twitter account by Woodward, showing an exposure of till(?) very similar to the one shown earlier, and again near the village of Tsepelovo. A prominent line of cobbles and boulders is plainly visible running through the center of the deposit, and lines of smaller fragments and some crude stratification are also discernable, while the lower foreground is littered with the same subrounded to rounded stones, as seen in the previous photo (fig. 4-9) from elsewhere in the same area.

The situation is much the same on Mt. Smolikas with ". . . thick diamicton deposits at an altitude of 1,000 m (3,200 ft.) a.s.l." and three stacked sub-glacial tills are defined. A fourth, late, or recent, phase has also been claimed (Leontaritis et al. 2020, pp 70-72). Three "discrete sedimentary packages . . . related to glacial activity in the uplands have been identified . . ." around Mt. Olympus (pp 72-73). The situation is, again, much the same on the Peloponnese mountains; three phases have again been identified but with less certainty, which does not surprise me.

These include heavily eroded diamicton deposits buried within an alluvial fan at an elevation of 950 m (3,100 ft.). Leontaritis et al. (2020, p. 75) suggest they may be interpreted as till deposited by a thin, long glacier draining the central ice field, which I would consider far-fetched at best. Heavily eroded diamicton deposits and alluvial fans bespeak water action, and "a thin, long glacier" seems a contrivance to supposedly explain an extensive or extended deposit of what these and all workers are interpreting as glacial tills, when, to me, they look more like they were water-shaped and deposited. The following is an example of the valley till/moraine/diamicton exposed in a road cutting on the flank of Mt. Chelmos, showing that it is much the same as the others we've seen on Mt. Tymphi, but maybe the boulders, stones, etc., are a bit more angular.

Fig. 4-11: Diamicton exposure on Mt. Chelmos.
Image: R. J. Pope et al., 2015.

The only other area of Greece considered to have been glaciated is the Sterea Hellas in central Greece. Here again, the same features are listed for the higher parts of the mountains, with glacial valleys and till reported from the flanks. Little information is provided by Leontaritis et al. (2020) or anyone else, due to the limited work that has been done in the area. However, it being so far south like others mentioned, I remain very doubtful of glaciation.

While I would allow ice fields and glaciers on the heights of the northern mountains, as all workers appear to agree, and, while reserving judgement on the lower valleys, we will move on to the other types of deposits mentioned; these are all agreed to be water-transported. The source and supply of these sediments is considered to be, for the most part, the debris generated by the glaciers and by any other agents operating, the main one being the freeze/thaw cycle which can, with regular cycling, rapidly erode certain rocks and generate large amounts of debris of all sizes, and sharply angular in shape. The glaciers must then bring this material off the heights and down into the valleys so that the meltwater streams can transport it to even lower levels.

On being transported by water, the debris, etc., becomes fluvio-glacial (or glacio-fluvial), which is to say, glacial material fluvially transported, and the amount and size of the debris moved depends almost entirely on the amount of water and its speed, while the initial shape of the fragments also plays a lesser part. And it is this water transport that ultimately rounds off the stones and boulders. See for example Krumbein (1941) or Domokos et al. (2014).

Logically, the water moves fastest on the upper and steeper slopes, and larger fragments, including boulders, are easily transported. This is commonly seen in many mountain torrents today, particularly during the spring snow-melt floods. The degree of rounding is proportional to the distance travelled, and as the valleys become gentler, the water slows down and the largest boulders are deposited, followed by progressively finer material, as is perfectly logical.

As the valley reaches low ground, it may blend gently into a more or less level floor and become a typical river valley, with the usual slow-moving river wending its way along somewhere near the center line. At this stage, the river's speed has greatly decreased compared with the heights, and, since its transporting abilities reduce in parallel, the coarser material is dropped and the river then remains only capable of transporting progressively finer material, all of which would be considered fluvial.

At the other extreme is a valley where the river transitions suddenly from a mountain front out directly onto a flat plain, with a sharp reduction in speed at that point. As the river leaves the confines of the narrow mountain valley, it is free to spread out in all directions and naturally loses most of its speed and its ability to transport. In this case, the river deposits most of what it's carrying right where it debouches from the mountains, leaving an alluvial fan, which takes much the same form as a river delta. (A similar situation is seen with mountains that border directly on the sea, as mentioned earlier).

Collectively, the workers on the Ice Age in Greece have identified glacial evidence on the high mountains of northern Greece, the Pindus in particular. Such evidence gets harder to come by as we move south. Certain questions were also raised by the extent of the glaciers shown on maps, their termini not all adhering to a common equilibrium line altitude (ELA). This is a major problem for the claims relating to glaciers made by a number of workers herein.

Physics demands that the ELA not vary by much over short distances, and hence, this evidence must have a different explanation. The nature of the tills with their roundy boulders is also problematic, as ice doesn't do this. I would, therefore, have to consider that, whether or not there actually was glaciation on Greece during the Ice Age, much of the evidence does not appear to be glacier-derived and must have another origin, the rounded shape of the stones suggesting water, of course.

We will now move on to study the rivers and valleys of Greece.

Chapter References

Brevik, E. C. & Reid, J. R., 2000, Differentiating Till and Debris Flow Deposits in Glacial Landscapes, *Soil Horizons*, Vol. 41, No. 3: 83-90 (p. 83).

Domokos G. et al., 2014, How River Rocks Round: Resolving the Shape-Size Paradox, *PLoS One*, Vol. 9, No. 2.

Evans, I. S., 2006, Glacier Distribution in the Alps: Statistical Modelling of Altitude and Aspect, *Geografiska Annaler,* Series A, Physical Geography, Vol. 88, No. 2, (Abstract).

Flint, R. F., 1971, *Glacial And Quaternary Geology*, John Wiley and Sons, New York (p. 141).

Hughes, P. D. et al., 2006, The glacial history of the Pindus Mountains, Greece, *Journal of Geology*, Vol. 114, 413-434.

Hughes, P. D., Woodward, J. C. & Gibbard, P. L., 2006, Quaternary glacial history of the Mediterranean mountains, *Progress in Physical Geography*, Vol. 30, Issue 3, p. 345.

Hughes, P. D., Gibbard, P. L. & Woodward, J. C., 2006, Middle Pleistocene glacier behavior in the Mediterranean: sedimentological evidence from the Pindus Mountains, Greece, *Jour. Geol. Soc. Lon.*, Vol. 163, pp 857–867.

Hughes, P. D., Gibbard, P. L. & Woodward, J. C., 2007, Geological controls on Pleistocene glaciation and cirque form in Greece, *Geomorphology*, Vol. 88, Issues 3–4, pp 242–253.

Krumbein, W. C., 1941, The Effects of Abrasion on the Size, Shape and Roundness of Rock Fragments, *The Journal of Geology*, Vol. 49, No. 5, (Jul.-Aug., 1941) pp 482-520.

Leontaritis, A. D., Kouli, K. & Pavlopoulos, K., 2020, The glacial history of Greece: a comprehensive review, *Mediterranean Geoscience Reviews*, Vol. 2, April 2020.

Nemec, W. & Postma, G., 1993, Quaternary alluvial fans in south-western Crete: sedimentation process and geomorphic evolution. *Special Publication of the International Association of Sedimentology*, Vol. 17, pp 256-276, cited in Hughes et al., 2006.

Ohmura, A. et al., 1992, Climate at the equilibrium line of glaciers, *Journal of Glaciology*, Vol. 38, No. 130, pp 397-398.

Pope, R. J., Hughes, P. D. & Skourtas, E., 2015, Glacial history of Mt. Chelmos, Peloponnesus, Greece, *Geol. Soc. Lon.*, Special Publication 433, 211-236, (Abstract).

Woodward, J. C. & Hughes, P. D., 2011, Glaciation in Greece: A New Record of Cold Stage Environments in the Mediterranean. In: *Developments in Quaternary Science*, Vol. 15, Amsterdam, The Netherlands, 2011, pp 175-198 (Chapter 15) pp 175, 189.

CHAPTER FIVE:

RIVER VALLEYS OF GREECE

FROM THE GLACIERS IN THE MOUNTAINS AND VAL-
leys, we naturally transition to the rivers that formerly flowed, and
those now flowing, in those previously, and supposedly, glaciated
valleys. Of all the great uniformitarian eroding, transporting, and
depositing agents that are operating in the world at all times, by far
the most work is ascribed to rivers. We met some fluvial references
in the last chapter, and so we will now move on to the deposits,
described as "alluvial."

The word *alluvial* is used in connection with any river or floodplain deposits, as in an "alluvial-plain" which is formed when a river
overflows its banks during regular flood events. Such plains are typically found in the lower reaches of a river, where the valley is wide
and the river slow. The typical deposits of flood plains are, as most
people are aware, generally comprised of the finest materials, such
as fine sand, silt, or mud, and these are also known more generally
as over-bank deposits.

The coarser and heavier material, such as sand, coarse sand,
and gravel, if any, remains in the river channel. That's because the
flow of water of any typical river today, even during a high flood, is
not powerful enough to lift this material out over the bank.

In fact, it often cannot move the larger material such as coarse sand or gravel very far down a river, and then only by "dragging" it along the bottom, comprising what is known as the "bed-load," and only sand, silt, and mud make it down to the lower reaches of a river or anywhere near the river's mouth, as noted earlier.

The foregoing is a very general description of the behavior of rivers, applicable globally, and is merely the logical mechanism by which meltwater and rain are conducted from the highlands, and general land surfaces, to the seas. Groundwater, of course, is continuously feeding rivers in the absence of rain or snow. The same process worked the same way during the Ice Age, except conditions at any locality are assumed to have been different from the present, to one degree or another.

In the case of Greece, which, as we've seen, didn't have much in the way of glaciers, but, whether we can assume more snow and rain or not, the conditions would not have been a great deal different from today. By this I mean that since Greece wasn't covered by an ice sheet or huge glaciers, the landscape could not have been modified by the action of glacial ice to any appreciable degree. Tectonic activity is certain; change in sea level is definite, and the Aegean basins have been filled with enormous quantities of sediments that appear to have come from the north-northwest, by some means.

The following photos show examples of the types of deposits mentioned in the text. It should be noted that each type can, and will, vary enormously in terms of sediment and structure, and thus many of these features can be difficult to identify any way positively as a definite "type" of deposit, or distinguish it from other types of similar-looking deposits, but which might have a different origin. The fact remains, however, that water, in one way or another, whether solid or liquid, is involved in virtually every sedimentary deposit, except eolian (wind).

Fig. 5-1: Fluvio-glacial sediments consisting of rounded gravel and sand with some layering apparent, Ontario, Canada. Image: Gov. of Canada, website Agriculture and Agri-Food Canada, Canada.ca.

Fig. 5-2: Fluvial (in-channel) deposits of fine gravel, sand and silt, in and beside the Morava River, Austria. Image: File # 557388123 © matuty, Adobe Stock, Standard License.

Fig. 5-3: Alluvial fans in Death Valley, Nevada, U.S., classic shape with multiple fans merging, with the fans dissected by braided stream channels. Image: USGS.

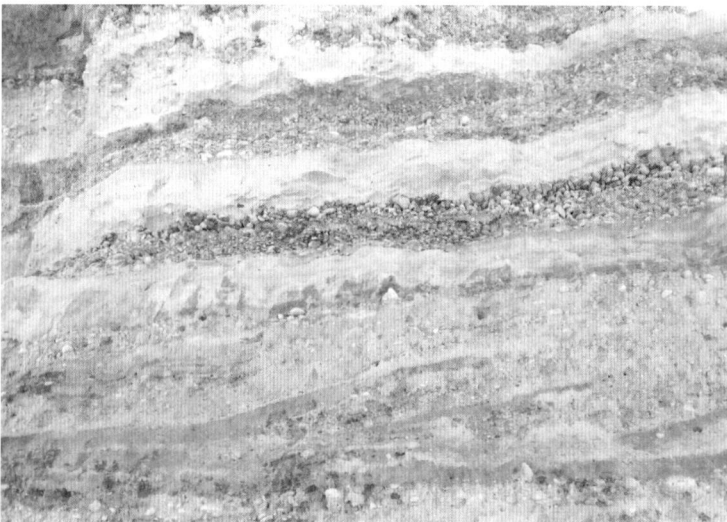

Fig. 5-4: Alluvial fan deposit showing repeating sequences, indicating multiple floods, Jordan River, Jordan. Image: Daniel J. Goode, USGS.

Fig. 5-5: Unconsolidated flash-flood deposits in a road-cutting beside the gorge east of Paleochora, southern Crete, Greece. These repeating units are several feet thick. Image: ID 6937444 © Paul Cowan, Dreamstime.com. Royalty-free Lic.

As mentioned, these sediment types and deposits can vary widely in any area under study, and their compositions are dependent on the source material and the topography, as well as the drainage patterns and so forth. In Greece, these deposits are found in valleys, at their mouths, in plains, at the seashore, in sedimentary basins, and elsewhere—and, in many cases, where they definitely should not be, i.e., the Aegean basins.

Mainland Greece, as it turns out, also includes a number of sedimentary basins that are, in fact, grabens or half-grabens exactly like the ones comprising much of the submerged Aegean area. These basins have also been filled with sediments, and today form levelish plains in various areas of Greece, as shown on the maps of fig. 5-6.

Most basins have rivers or streams flowing through them, and are, in essence, simply river basins, gently sloping much like any river basins anywhere, varying between wide level plains and narrow river valleys. Needless to say, both basin formation and infilling are considered by orthodox geology to have taken the last 10 my or so, the Neogene to Quaternary periods, to get them to their present condition, as indicated by the caption for map (a).

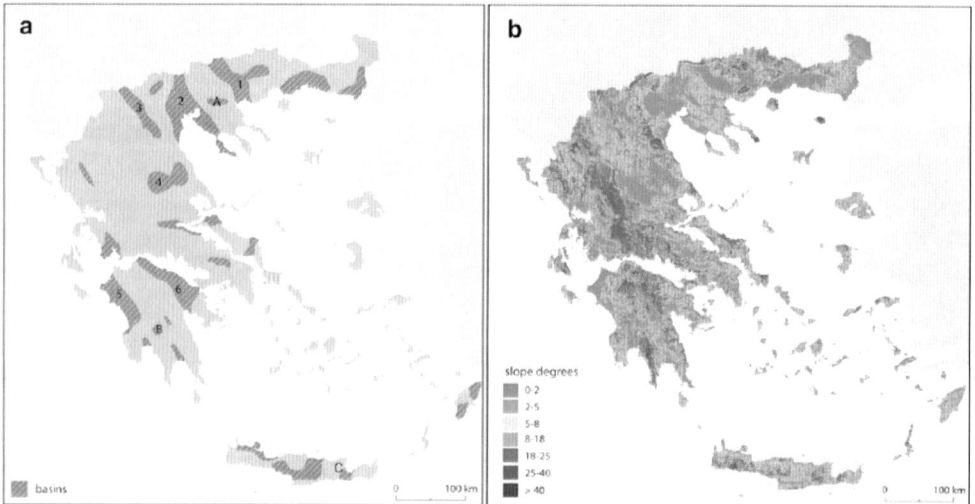

Fig. 5-6: (a) The main Neogene-Quaternary sedimentary basins of Greece, where the bulk of Pleistocene sediments are located (b) Slope map of Greece showing flatter areas comprising basins, valleys and coastal areas. The basins shown on the left map can be correlated with the larger areas on the map on the right. Image: Tourloukis & Karkanas (2012) modified after Mountrakis, 1985.

The map above is from a paper (Tourloukis & Karkanas, 2012) on the preservation of archaeological artifacts in Greece. As part of the discussion, the authors describe what is thought to have been the general erosional/depositional environment of Greece from the

Pleistocene Epoch to the recent/present. On page 7: "The dominant influence on erosion and sedimentation is 'precipitation input' particularly as a result of extreme events, such as large-scale floods or deluges (sic) generating high turbulent surface runoff."

Apart from the fact that *deluge* is not a word one hears very often from the orthodox geological community, the writers offer no justification for this suggestion, other than unexplained changes in climate, rainfall, snowfall, or changes in the frequency of any of their ". . . short-lived, high-magnitude flood events . . ." all of which would, of course, massively increase river flow, erosion, transport, and deposition.

Such events (sounding suspiciously similar to the "alluviation events" of van Andel and Zangger, from earlier) would be extremely useful if they did, in fact, occur, especially when one is trying to explain the presence of coarse clastic material where uniformitarian geology says such material has no right to be, even as a result of the worst flooding to occur in Greece today. As the writers state later in the paper, the main erosional episodes going on in Greece today are landslide events due to the rugged relief and steep slopes of the mountains, something to which the lack of plant cover very clearly contributes. So, clearly, there's not much geological activity going on in Greece today.

In discussing the structure of Greece in general, the writers tell us that the vast majority of Pleistocene (Ice Age) sediments are buried in the basins, and they tell us that these basins have been subsiding for the last 5 my, and right up to recent times. (We'll ignore for now the fact that we know they weren't.) Hence, they have been acting as sediment receivers, as they have always formed the low ground to and through which rain and rivers have flowed. However, they further state that these basins are "low-gradient" as can be seen on the slope map shown in fig. 5-5, which is in keeping with the their being depositional environments. It also means that any rivers in those basins flow slowly.

Interestingly, they tell us that significant changes took place during the early to mid-Pleistocene, i.e., changes in drainage patterns, rapid rifting in the Gulf of Corinth, uplift of Peloponnese basins and a change in direction of tectonic extension related to a major "kinematic transition" affecting the entire eastern Mediterranean area. Early to mid-Pleistocene is very recent, during the Ice Age supposedly, and we recall that Neumayr gave a date of 1 my for these events, which would be mid-Pleistocene. However, we must not lose sight of the fact that the duration of the Ice Age, and the Pleistocene epoch itself, are very debatable, and these tectonic events may have occurred a lot more recently than orthodox geology presumes, and as we saw in chapter 1 herein earlier.

This suggests that some major event of a tectonic nature occurred not too long ago and harkens back to the back-and-forth motion of the African/Aegean tectonic plates we saw earlier in this study, which involved complicated movements and very prominent stretching and rifting. Tourloukis & Karkanas (2012) also tell us that during this phase, sediments in many of the newly uplifted basins (grabens) were eroded away, without saying how, exactly. In any case, this supposedly explains the absence of archaeological remains pertaining to the mid-Pleistocene.

The writers refer to this phase of tectonism using another word one does not hear very often from a modern geologist—*paroxysmic*, which, of course, means "violently catastrophic," as the "deluges" referred to earlier would also suggest (Tourloukis & Karkanas, 2012). Given the reference to violent events, it would seem natural to suppose that such events might have had something to do with eroding the sediments of the basins.

Steep slopes and high-gradient valleys are the norm in the Greek mountains today, and thus material is easily moved from the heights all the way to the lowlands, or at least it would be if there was plenty of water. However, the rivers of Greece are now all relatively small, especially near their sources in the hills, and many are ephemeral, and so the amount of material brought down is extremely limited and generally fine-grained, very different from the coarse, glacio-fluvial material of the Ice Age.

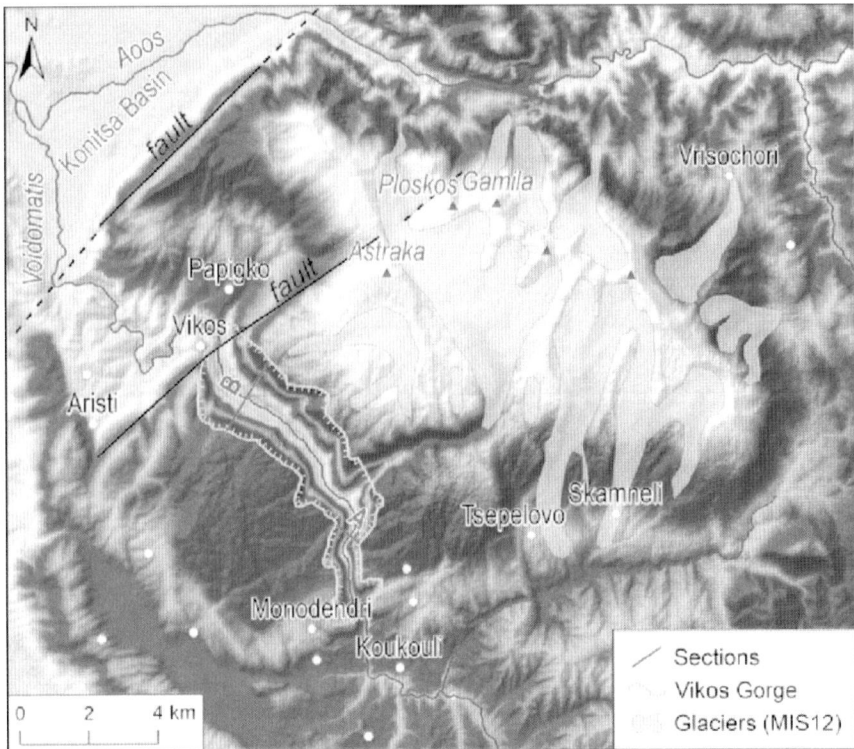

Fig. 5-7: Map of Mt. Tymphi area showing rivers Voidomatis and Aoos, and the Vikos Gorge. The rivers are seen to join in the Konitsa Basin. Image: Telbisz et al., 2019 (glacier from Hughes et al., 2007).

As we saw earlier, the major focus of work on the Ice Age in Greece was the Pindus mountains, and in particular, Mt. Tymphi. Logically following on from that work were studies of the valleys of the rivers that drain the area, the Voidomatis and the Aoos, as shown in fig. 5-7 (repeat of fig. 4-2, from previous chapter).

A lot of work has been done on the Voidomatis river, in particular, much of it by the same people who studied the glaciation of Mt. Tymphi—Woodward especially, as well as some other British and many Greek geologists. We will begin with the Voidomatis River, which drains Mt. Tymphi on the south and west sides, as it is representative of many Greek rivers and basins.

As we learned from the glacial studies, the heights and upper valleys of Mt. Tymphi retain large amounts of glacially derived sediments, identified by various workers as glacial tills, moraines, and so on, and often simply referred to as diamictons. These comprise the sediments most likely to be transported by the river, along with whatever it picks up, or falls into it from the surrounding hillsides, which, according to Woodward et al. (1992, p. 209), is mostly just fine sediment washed off by rain, such as is common to most rivers in the world today.

The river can be divided into three sections, the higher parts with a number of tributaries flowing through the upper basin, the mid-section flowing through the Vikos Gorge, and the lower section exiting the gorge and mountains and flowing across the Konitsa basin, a flat alluvial plain, where it joins with the Aoos. The map of the river basin in fig. 5-8 shows the main features and localities referred to in the following text.

Woodward and other British workers published a number of papers, in the 1980s and 1990s, on both the glacial history of Mt. Tymphi and the depositional history of the Voidomatis river, and they collated much of this work in a paper in 2008 on the connection between them. In the early papers, they report identifying four distinct sedimentary units, three older Pleistocene units, and one recent Holocene, unit (Bailey et al., 1989, pp 145-146; Woodward et al., 1992, pp 209-211).

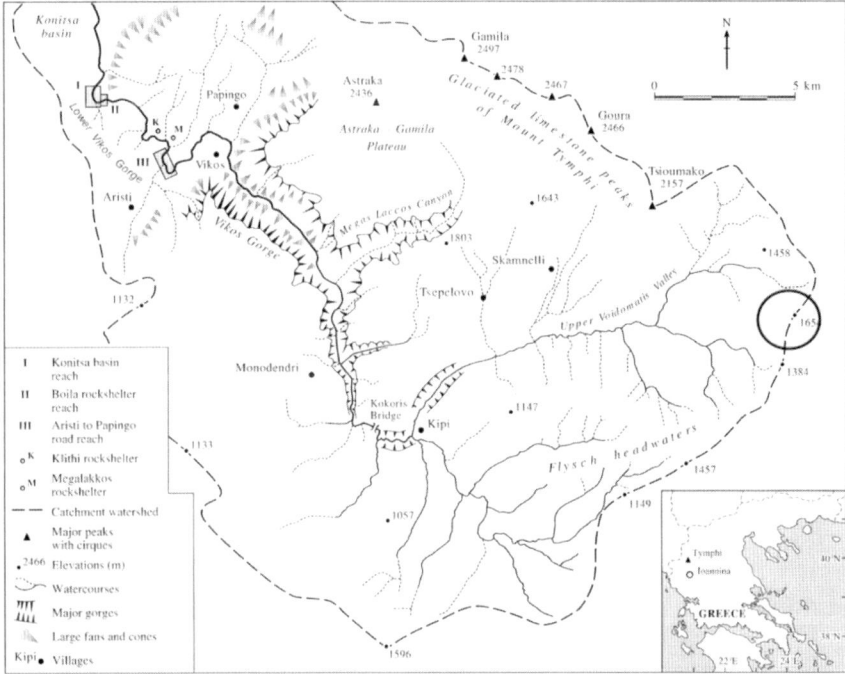

Fig. 5-8: Map showing Voidomatis River basin, its catchment watershed and main features and locations. Image: Woodward et al., 2008, p. 45.

These units are found on the banks and slopes adjacent to the river, and form terraces as infillings of hollows, embayments, and so forth, along its course. They are essentially stacked up against the sides of the valley, while the river flows below and through them along the valley.

The oldest is called the Kipi unit, and is found just upriver from the Vikos gorge, exposed at one location only, the Kokoris bridge, see map above and (B) in fig. 5-12. The unit consists of gravel and sand, etc., but since it contains only a little limestone, it is thought to derive from the area further to the northeast, beyond, that is, *outside* the Voidomatis valley and close to where the Aoos originates.

This is the main area where one particular and dominant type of rock debris the Kipi unit contains is exposed (ophiolite, an igneous-type rock). In the Voidomatis basin itself, an exposure of this rock occurs only in a very limited area at the top of the ridge, at the extreme east of the Voidomatis catchment basin, and is shown circled in black in fig. 5-8. The dashed perimeter line is the limit of the catch-basin of the Voidomatis, the watershed, indicating that the outcrop is at the top of a dividing ridge.

As can be seen, it is not an extensive outcrop; it's well-away from even the sources of the tributary streams of the Voidomatis and, being at the very top of a ridge, it couldn't have received much rain wash. While it may have suffered frost shattering at some point in the past, if not at present, there could not have been enough rain over so small an area to wash much of that material down to the small tributary streams, which themselves could not have received much rainfall either, to swell them into the raging torrents required for such work.

Since nothing is going on here at present, it is difficult to see how fragments of this rock made it down to the Kokoris bridge in sufficient quantities so as to constitute the "dominant" rock type of the unit. Or at least it's difficult under anything like present conditions. Even in the past some agent was required, water being the usual one. We can't resort to ice-transport since no glacier could ever possibly have existed anywhere near the source of those ophiolites at the top of the dividing ridge. Apart from that, it's difficult in the first place to see how the exposed ophiolite rock could have been fragmented to any great extent. If it was covered in the snow of the Ice Age, frost action couldn't have accessed it, and, as noted above, rain could have had little effect.

Fig. 5-9: Upper reaches of the Voidomatis showing the sedimentary debris lying in and about its bed, temporarily dry, presumably. These sub-rounded boulders are rather large for any river to move, let alone the upper reaches of the Voidomatis as it is today. As Macklin & Woodward suggest in reference to this image, the climate must have changed considerably. Image: Macklin & Woodward (2009).

In today's climate, which is cold and wet in winter, with occasional heavy rainfall, frost and snow, there is no shortage of weathering agents and, while the lower slopes do have some tree and plant cover, the upper slopes are generally bare, with little such cover and always exposed to the elements. The main rocks of the valley are hard resistant limestone and soft and friable flysch (thin-bedded silt and sandstones). Woodward et al. (1992) tell us that the flysch slopes: ". . . are highly susceptible to erosion and provide an important source of easily erodible and transportable *fine* sediment" (author's italics). Both types of rock, however, do provide some fragments and detritus, i.e., coarse material.

Fig. 5-10: Voidomatis River valley, Vikos Gorge, Pindus Mountains, Greece. Bare cliffs in the upper reaches with plant cover on the lower. Image: CC Attribution-Share Alike 4.0 International license. Photo: anas.dimitris ID GR2130001.

In the limestone areas, most rainwater drains into the karst surface and little reaches the river, so that today, and at earlier warmer times, very little limestone material reaches the river, other than by falling in directly from its banks or into its tributary streams. The nature of the valley is shown in fig. 5-10, and despite the high degree of exposure, cliffs and steep slopes, present-day erosion and rain wash supplies only a limited amount of fine sediment to the river, which must make one wonder as to what agent was acting in the past, and how energetically.

Hence, and yet again, present conditions, and currently operating agents and processes, do not seem to be capable of doing the work that has obviously been done in the past.

Fig. 5-11: Boulders in the Voidomatis river. They are suffering some erosion/dissolution but the moss on the upper parts of some tell us that these boulders haven't moved for quite the while. Image: ID 154484709 © Kordoz, Dreamstime.com. Royalty-free license.

Fig. 5-12: These boulders and plants in the nearby Aoos River tell the same story. Image: ID 158554523 © Albertoloyo. Royalty-free License.

For example, fig. 5-11 shows a section of the Voidomatis river as it looks today, its bed strewn with large boulders that the present-day river, even at peak flood, could not possibly move. The moss and algae on the boulders, and the trees and plants by the river, tell us that very little erosion or transportation is going on here at present, apart from some dissolution of the boulders.

The boulders in the photographs (figs. 5-11, 5-12) are seen piled one atop the other to a height of a few meters above the level of the river, and, judging by the size of the bushes and trees, are up to a meter or more in diameter. Since these boulders have moss and algae on some of them, and even the small ones are obviously far too large to be moved by the present river, even in flood, and since they're all rounded, they must have been transported and deposited by a much more powerful and fast-moving body of water, i.e., a much bigger river or a much larger flood. Further, in general, the presence of moss, algae, slime and so on, on river rocks is a dead giveaway that that river is not doing any eroding.

The circumstances pertaining to the origin and deposition of the (Kipi) unit are difficult to explain by conventional means, Ice Age or no. Conditions must have been different and a different agent must have been involved, as compared to either today or during said Ice Age. Woodward et al. (1992, p. 211) suggest that the unit's presence indicates a different drainage pattern in the upper reaches of the river, at the time the unit was deposited, followed by the Aoos somehow *capturing* the former headwaters of an early version of the Voidomatis to create the present drainage pattern.

As can be clearly seen in fig. 5-7, ridges and mountains divide the *present-day* basin of the Voidomatis from that of the Aoos, and the entire basin of the Voidomatis is defined by the same kinds of ridges that separate its basin from that of the Aoos on the other side of the dividing ridge, while the river Aoos extends back up along the eastern side of the ridge, at the top of which the only ophiolite rocks outcrop in the Voidomatis basin. The *river capture* referred to here by Woodward et al. (1992) is an old (and venerable) concept that was dreamt up in the mid-19th century to explain awkward situations such as these. We will examine it shortly.

The Aoos river does drain an extensive area containing ophiolites, lying to the east of the Voidomatis, i.e., on the other side of the ridges (fig. 5-7) that define the catchment basin of the river. Apart from the small outcrop on the ridge itself, in its own basin, that is the only other apparent source of the ophiolite clasts found in the Voidomatis deposits.

There is, however, another very extensive area of ophiolites further to the north, spanning the Greek-North Macedonian border. This, of course, would entail eroding rock fragments off in the north, transporting them many miles south, then lifting them up and over the ridges separating the two basins, and sweeping them down into the Voidomatis valley. This, of course, would require some powerful agent moving from the north—and ice wasn't it.

The Kipi unit lies at the surprising elevation of 56 m (180 ft.) above the river. How it got to where it is now, so high above the level of the river, the writers do not venture to guess. Since glaciers didn't make it this far down the valley, it had to be water, carrying a load of coarse debris, that got it up that high somehow. It couldn't have been put there when the valley was shallower, i.e., before it was "carved out," since eroded fragments of the limestone rock of the valley would have been unavoidably included in the deposit.

This lack of limestone is conveniently explained by claiming that this deposit was laid down before the Ice Age ever affected Mt. Tymphi (Woodward et al., 2008), which actually explains nothing and simply avoids the issue, obviously enough.

The Ice Age supposedly began over two million years ago, but since they give an age of 150,000 years, or so, to the ophiolite clasts in the material, one wonders how Greece held out for so long against the cold. The authors offer no explanation for the ophiolites, but only suggest a derivation from the areas on the other side of the ridges, to the east, via ". . . a fluvial system with a much more extensive catchment . . ."

This would mean that the entire general land surface of the region was formerly at a much higher elevation, higher even than the level of the separating ridges, which supposedly didn't exist in their current form. This elevated plateau must also have predated the valleys in order that some ancestral Voidomatis river could make its way, in some former valley, from the ophiolite area to the east of the present dividing ridge, and thence into the Voidomatis valley, which somehow must have pre-existed. Woodward et al. (2008) write off this little problem by declaring it pre-glacial and therefore not relevant to glacio-fluvial linkages.

Ignoring the problem does not help, nor does inventing spurious "ancient river systems" that could not possibly have existed. At the same time, an "ancient pre-glacial river system" even if it could have existed where they suggest, would likely be no more powerful than a river today, since the pre-glacial climate would presumably have been warm and somewhat similar to that of today. This, in turn, means that conditions of erosion, transportation, and deposition would also be little different from those of today, and as we can see and as we're told (Woodward et al., 1992), very little in terms of erosion and deposition is going on in this valley at present. And "very little going on" is a feature common to approximately every river valley on this planet at any given time, and nowhere do we see any "carving" processes in action.

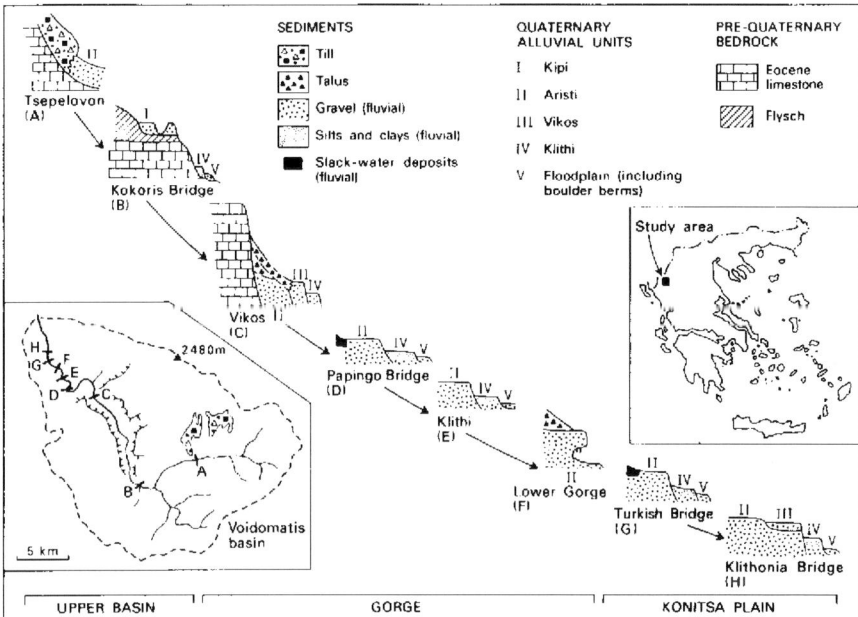

Fig. 5-13: Schematic cross-section of the Voidomatis valley showing the different depositional units and the terraces they form. Image: Bailey et al. 1990. As can be seen in the profiles of the terraces, the Kipi Unit sits on a Flysch ledge above the Kokoris Bridge, while Unit V, at the bottom right at most localities shown, is the latest and forms today's floodplain. The other three units mostly lie stacked against, rather than on top of each other, and this is conventionally interpreted as alternating periods of deposition and erosion to form the terrace series. Image: Woodward et al., 2008.

Hence, "river capture of the former headwaters of an early version of the Voidomatis" can hardly explain it. The presence of the Kipi unit and its contents where we now find them seems to suggest that either the entire Voidomatis valley, and the other surrounding ones, were once a lot shallower, and have since been greatly deepened, or else water, well over 200 ft. deep, once swept down this valley, from the east or, more likely, from the north. The latter, at least, would account for all those large boulders in the rivers.

Before we move on, a note on "river capture" is required.

River Capture

There are, generally speaking, two types of river capture: the captures we know occur in the present day because we have witnessed them or have examined the evidence of their recent occurrence, and the ones that are theorized to have occurred in the distant past. The two are completely different from one another.

The first type occurs as a result of changes in the landscape due primarily to humans. Also, past tectonic movements such as elevation of the ground and changes of slope and direction can also be discerned and identified as the cause of diversion, and these are often evidenced by an abandoned river channel and are thus easy to account for.

Landslides and volcanic eruptions have also caused rivers to be diverted; earthquakes sometimes cause a change in the terrain or possibly a chasm, causing the river to change direction or be diverted, sometimes into another river's channel. Glacial retreat over a differently oriented topography and/or glacial sediment deposition may divert a meltwater stream, and an advancing glacier may block a stream and change its course.

In all these examples, diversion is the main reason for capture, but such diversion does not necessarily lead to capture, and it more usually results in the stream simply following a new course of its own. Such types of capture or diversion are easily identified from the field evidence, and little mystery attaches to them.

It is an entirely different matter, however, with the other type of strictly stream capture, or so-called piracy; this is the type mentioned by Woodward et al. (2008) above, with reference to the Voidomatis/ Aoos river system. *The Encyclopedia of Earth Sciences,* "Stream Capture, Piracy," defines stream capture as occurring when: "an actively eroding low-level stream encroaches on the drainage of a nearby stream flowing at a higher level and diverts part of the water of the higher stream.

It may be caused by *abstraction* (merging of two valleys) *headward erosion* (up-valley erosion, which means the river is eroding backwards, uphill, toward its source) *lateral planation* (sideways erosion) or (self-explanatory) *subterranean diversion.*" "Capture" and "piracy" are now essentially synonymous."

The entry continues: "The "winning" stream may be called the *captor, diverter* or *predatory stream*" See Lauder, W. R. (1968). (It's always good, it seems, to foist a Darwinian analogy on the reader at every possible opportunity, and apply it to anything, no matter how inapplicable. Are we required to think the streams are [sub] consciously hunting and fighting each other in their competition for survival? And, no doubt, the "winning" stream is clearly the "fittest" to survive). [italics in article; this author's irritation in parentheses]

Even though the above definition clearly states an *actively eroding* stream, such a thing, and the goal of stream capture that it supposedly achieves, has never been observed to actually be in action in the present day, like many of the other geological processes we've discussed in these pages.

As Paul Bishop, in a 1995 paper, tells us: "The processes involved in drainage rearrangement are not as self-evident as its abundant literature indicates. This is especially the case with the commonly invoked *stream capture.* The key process in stream capture, namely, drainage head retreat is difficult to envisage as a normal part of drainage net evolution" . . . "Stream capture may therefore be a relatively rare event in drainage net evolution. This, and uncertainties with interpretations of elbows of capture, mean that stream capture should not be routinely invoked in interpretations of long-term drainage evolution" (Bishop, P., 1995). Again, we see that great overachiever "evolution" being invoked.

What is not said is that supposed "headward erosion" occurs at the very point where the stream is at its absolute weakest, as we'll see, and it must be said that Woodward et al. (1992, 2008) were rather casual in their invoking of capture of the "former headwaters" of an early and more extensive version of the Voidomatis. This concept, or theory, of river capture has its origins back in the 1860s, called up, or maybe dreamt up, to explain those same "elbows of capture" that Bishop tells us are beset about with uncertainties.

The theory of river capture, and especially "elbows of capture" will be dealt with as part of our analysis of the science of geology in a later installment, but for now, and with reference to the Voidomatis/Aoos system, it is fairly obvious that Woodward et al. are simply avoiding the issue, hence their declaration of its being pre-glacial, and not relevant to the topic at hand.

As Bishop clearly stated, it is "headward erosion" that is mostly invoked in the context of stream capture, thereby falling back on 19th century convention. And this is often done for the same reasons as indicated by Woodward et al. in the Voidomatis, foreign sediments in a river basin that have no apparent mechanism of transport into that basin. Other reasons include flora and fauna from a different river in the absence of any connection between the two.

Elbows of capture are simply right angle (or acute) bends in a river that are explained as the two rivers intersecting at right angles, typically where one somehow takes over as the main river. Right angle bends, and especially acute ones, are considered as not the normal result of general landscape erosion, because rivers tend to follow a more linear, if irregular and sinuous course, and many drainage nets are convergent. Hence these sharp bends are contrary to what are considered normal river behavior and thus, need a special explanation.

In the case of the Voidomatis, the implication is that the general land surface was much higher in former times, higher than the ridges now separating the two river valleys. Back then, a much more extensive Voidomatis drained an upland (plateau-type) area to the east of its present valley, and by this means somehow eroded, picked up and transported all the ophiolite material from the east and deposited it as the Kipi Unit in the present, or former version of the present, Voidomatis valley.

Later, the Aoos, which was supposedly lower than the Voidomatis, somehow, never explained, started carving out its own valley, backwards up into the highlands until such time as it had, again somehow, eroded its way to the "headwaters of a former Voidomatis," which were flowing west across the plateau. The Aoos then undermined those headwaters by cutting out a valley through the plateau and creating a ridge separating the valley of the Aoos from that of the Voidomatis, as we now see on the map, and especially in fig. 5-7.

The plateau itself was subsequently, and somehow, eroded into the uneven mountainous landscape we see today, by agents typically assumed to be rain, rivers, and frost, with all of this occurring prior to the Ice Age. However, prior to the Ice Age, the climate should have been similar to that of today. Since, according to Woodward et al., only small amounts of fine sediment are now brought down by the Voidomatis, little erosion can be going on. Thus, we may ask what "predatory" agent did all that valley excavating back in those formerly complacent times?

The conclusion one is forced to is that river capture is another invented process, especially since a wider review of the literature reveals expressions along the same lines as Paul Bishop's. As Stokes et al. (2018) of the Massachusetts Institute of Technology state in the very first sentence of the abstract of their paper on river capture in the Amazon: "River capture is thought to trigger abrupt changes in evolving continental drainage systems, but it is almost always inferred rather than observed, and the mechanisms that lead to capture are unclear."

Despite this admission, the authors immediately go on to describe how the Amazon River is now actually in the process of capturing the upper Orinoco. The Orinoco splits far up in its course to send a quarter of its water through a channel, called the Casiquiare, and on to the River Negro, a tributary of the Amazon, navigably linking the Orinoco to the Amazon. It was first reported in 1639 by Fra Acuña (though de Orellana [Spanish conquistador] reported discovering its confluence with the Amazon in 1541, during his famous voyage down that river) and later by others, while von Humboldt explored it in the early 19th century.

While Stokes et al. here claim that such diversions "can be temporary features," they don't say why. Instead, they claim that this is an example of an actual, ongoing, permanent river capture, and that the whole of the upper Orinoco will "eventually" flow through the Casiquiare and thence to the Amazon.

Later in the paper, they tell us that past river captures are recognized by "circumstantial evidence" and can only mention one "documented case of a rare example of an active capture," which, it turns out, is merely a meltwater stream diversion related to a currently retreating glacier (the Kaskawulsh, Canada, Shugar et al., 2017).

Obviously, if a glacier retreats over a ridge, rise, or some similar feature, the water can no longer flow on that side of the landform but must instead flow on the other side of the landform, presumably a ridge or high ground of some kind, that constitutes the drainage divide. This is no more than the same type of diversion mentioned earlier in connection with glacier retreat, and hardly constitutes "capture."

The foregoing is intended to show the very theoretical nature of "river capture." It cannot be summed up better than in a statement from Chinese scientists Niannian Fan et al. (2018): "While river capture occurs regularly in numerical models, field observations are rare." This shows again that computer models are the best thing to ever happen to the science of geology. Anything is possible in the world of computer graphics and the geological imagination, though it must also be said that when used with solid evidence, valid assumptions, scientific laws, and sound scientific principles, they are indeed a boon to geology.

Despite the complete lack of any kind of evidence for any kind of supposedly ongoing process, leading to, let alone actually accomplishing, river capture, we have a paper from Willet et al. (2014) that gives a listing of all the processes that lead to drainage reorganization, including river diversion and capture, and the (Darwinian?) evolution of river networks. They tell us that: "Dynamic aspects of these networks include channels that shift laterally or expand upstream, ridges that migrate across Earth's surface, and river capture 'events' . . . processes result in a constantly changing map of the network . . ."

The "dynamic" phenomena listed here are not actually seen to be going on anywhere at present, and they are clearly occurring only in the writers' imaginations or their uniformitarian computer models. The idea of "ridges migrating across the earth's surface" is definitely one of the most absurd I've ever heard; I have heard of sandbars and dunes moving in a small way, but that is hardly the same thing as "ridges migrating across the earth's surface." I would also challenge these writers to provide just *one* example of a "*constantly* changing map of a river network."

Just about every paper on the topic opens with a reference to W. M. Davis (1899) the man who wrote the bible on river morphology (The Geographical Cycle) and who, despite the earlier rejection of his concepts, still, or actually again, commands an apparently inordinate loyalty from the academic community.

Hardly a writer passes up the opportunity to reference him in any paper's opening pages, and this comes across as more of a preliminary declaration of dutiful fealty to the ruling, uniformitarian paradigm than anything else. This, of course, lets the reader know immediately that, despite the problems with river capture, the paper will follow convention and not contain anything radical, nor likely much in the way of new ideas.

I have only referenced four papers on this topic, but most of those consulted are much the same. The overall impression one gets is: despite an almost complete lack of evidence, no eyewitness observations, and no active mechanism shown to be effective, there remains a full and firm belief in the entirely theoretical concept of "river capture," as opposed to the definite "diversion" to be seen everywhere.

Yet again, we see that the origins of present-day conditions cannot be related to any process seen to be in action in the aforesaid present day, but recourse must, yet again, be made to the imaginary processes of W. M. Davis having acted in the past, when a simple "event" or disruption of existing conditions will do the job every time. This last statement certainly does defy precious uniformitarianism but the reports suggest that the correct answer is other than that given by orthodox academia.

To my mind, the most realistic, if not quite appropriate, analogy one can make for any river is with the rules of Carl von Clausewitz, the 19th century German military theorist, who, with regard to military campaigns, said: "Changes to the original plan should only be made under conditions of extreme necessity" (von Clausewitz, *On War*, 1834). (That "extreme necessity," mind you, often arrived with the first shot fired.)

A river, however, obeys von Clausewitz; from when it first begins to flow, a river will always flow along the thalweg of its valley. ("Thalweg" is the German word for the "way of the valley.") Since the thalweg of a valley is the line connecting the lowest points along the length of the valley, the river is perfectly obeying the law of gravity—like any liquid, always finding its lowest level. At any given point, therefore, along the course of a river, the water is at its lowest possible level and can, obviously, only flow from there toward an even lower level, which is the next point along the thalweg of the valley.

To flow in any other direction or course, therefore, will require that the river flow in a direction other than along the thalweg of the valley, thereby defying the law of gravity, which it cannot do. It is tantamount to flowing uphill, regardless of how little uphill it actually might be, and water, obviously, cannot flow uphill. Therefore, a river will not, by any imagined "lateral erosion" or by any invented "predatory capture," or by any other invented process, deviate from its original course.

This is demonstrated by every present-day river, and unless some event, or agent, which, by "extreme necessity," *forces* some change on it, that river will not budge out of its course, as I will prove at the appropriate time, if, that is, the foregoing logic of physics, the law of gravity, and the well-known behavior of liquid water are collectively insufficient.

The whole subject of river behavior, which is of the utmost importance, will be treated thoroughly in a later volume, where it will be examined in the context of geomorphology. Suffice it to say for now that the "capture" of rivers or of "former headwaters" has nothing whatever to do with the formation of the Voidomatis or Aoos river valleys, and the problem of the ophiolites where, on the face of it, no ophiolites should be found. To return to the subject at hand:

The next, and dominant, unit is called the Aristi, and this is the major alluvial infill of the entire valley and basin. Limestone is the main component of this unit, comprising boulders, cobbles, gravels, and a large amount of fine material, and which likely derive from the heights of Mt. Tymphi itself and presumably also from the (at least supposed) glacial moraines and other deposits, such as the tills, etc.

According to Woodward et al. (1992), this unit is the terraced remnant of an extensive aggrading (building up or depositing) pro-glacial river system. By this they mean that the river first filled the valley to the level of the terrace and then some other, necessarily more powerful, version of the river came along and swept most of it away again and off down the valley. The infill is up to 25.9 m (85 ft.) above the river—that's a lot of coarse, bouldery sediment to put in place and remove. See fig. 5-10 above, photo taken upriver from the Vikos gorge.

The Aristi unit has been divided into four sub-units, implying four phases of deposition. However, considering the glaciation of Greece was simple mountain glaciation, then one would expect to see the results of a yearly meltwater flood in the form of lots of thin layers of sand and silt and maybe some gravel, etc. We do not; we see only a few large units, all containing boulders and coarse sediments.

The next unit is the Vikos, which is a relatively minor one with a low volume and limited extent. It does, however, contain a significant proportion of ophiolite clasts, which begs the question of their origin, since the source is very particular. In this case, however, it is thought to derive from the reworking of sediment of the Kipi Unit, which would be a reasonable interpretation, were it not for the elevation at which the Kipi unit lies.

This would mean that some flood or other would have had to fill the valley so that it could access the Kipi unit, and erode, or to be more accurate about it, retransport the material down the valley. While this may be the case, the fact that the Kipi Unit lies at a height of 56 m (180 ft.) above the river makes one wonder, and have serious doubts.

The latest unit, the Klithi, not counting recent deposits of the river, consists of coarse sandy gravels, overlain by a distinctive fine-grained unit up to 2.5 m (8 ft.) thick. This unit also extends over the level-ish Konitsa Plain to the west through which the river flows on leaving the mountains. The Konitsa floodplain deposits are the same kind of coarse alluvial gravels (Woodward et al. 1992) as at the old Klithonia bridge, interpreted by various authors as alluvial deposits produced by a "sediment-laden, fast-moving, meandering river." The sands and silts are considered overbank flood deposits, while the coarse gravels are in-channel deposits.

As Lewin et al. (1991) state, the coarse sediments were (supposedly, that is) deposited by: ". . . an aggrading, meandering river system with suspended sediment concentrations in flood waters significantly higher than those of the present day." One must wonder why there would be more suspended sediment than the present-day; floods today can be no different from past ones; water is water and its transporting power could not be different. In the Konitsa Plain, the present land surface lies between 5 and 10 m (16-32 ft.) above the level of the present river bed. That height difference represents a lot of deposition by a lot of floodwaters, all deriving from the Voidomatis valley.

Of interest here is the suggestion of a "sediment-laden, fast-moving, meandering river," which is called on to explain the obvious flood deposits that cover the entire plain. Now, in order to transport coarse sediment, any river must be moving quickly, and with a large water volume—a flood, obviously.

However, fast-flowing rivers, in flood or not, do not meander; they always flow more or less straight down their valleys, or in a slightly sinuous course, and only fast-moving rivers can be "sediment-laden." Conversely, only very sluggish rivers, flowing in near-level floodplains, such as the Mississippi or Amazon as familiar examples, truly meander as that term is meant, and these rivers do not flow quickly, nor do they carry coarse sediment.

The trouble is that this layer of coarse sediments covering the Konitsa plain to a thickness of from 2.5 m to 7.5 m (minus the top-most 2.5 m layer of finer sediments) can best be explained by invoking a large, widespread flood that covered the plain in one sweeping motion and deposited all those coarse sediments as it did so. And that flood must have swept down and out of the Voidomatis valley. This, the most obvious of explanations, and the one that that obeys the laws of fluid dynamics vis-à-vis coarse gravels, cannot, of course, be suggested. Obviously, there is a relationship between the speed of water and the size of material that that speed can move, long since scientifically defined by 19th century physicists, as we'll see.

Instead, the official reason that coarse gravels cover the plain is that, as this fast-moving river flowed across the plain, it also (theoretically) meandered progressively from one side of it to the other. We touched on this earlier with reference to the paper by Willet et al. (2014), in which the writers discuss "channels shifting laterally," a process never seen except in the event of a mature river breaking through its levee on a flat floodplain and finding a new course to the sea; normally a river shrinks back into its old course after the flood.

This only has the appearance of "shifting laterally"; it is actually only a course-diversion due to the fact that the river swept its levee completely away and can't regain its old channel.

This occurs on the rare occasion that a river has been depositing so much material in its bed, and on its banks, for so long, that the river has raised itself, its levees, and the level of its bed completely above the level of the surrounding floodplain, often many feet or meters above it, such that the river is now flowing along the top of a narrow linear plateau above the floodplain. In which case, on breaching the levee, the river will now be flowing at a level below its old channel, and hence, and obviously due to gravity, cannot ever flow in its original channel again.

In fact, such occurrences are mostly due to local farmers building up the levees to contain the river. Any such river will not spontaneously deviate from its original line of march unless some "event" like a levee breach *forces* it to seek a new course for itself. This has happened with rivers throughout history, always suddenly, drastically, and of necessity, and almost always due to human interference. A good example is the River Po in Italy; excessive build-up of the levees by the locals to control flooding resulted in a levee breach triggered by an earthquake in 1570, and a new course for the river.

The obvious exception would, at first glance only, appear to be meandering rivers. If studied, however, it will be found that the river only meanders within certain limits, such limits being defined by the outer curve of the meanders on either side, by lines roughly tangential to those curves. Within these limits is the "meander belt" or "meander zone." This is the zone within which channel migration occurs. Its limits do not change or migrate across the valley, and the river's meandering course remains within them, the river itself always maintaining the same general line of march, as defined by the course here taken by the *meander belt*, regardless of what actual course the river takes within it.

In any meandering river, there is a fixed ratio between the width of the river channel and the wavelength of the meander, the latter being 10-14 times the channel width, while the radius of curvature of the meander is about 2-3 times the channel width, and these two values can change only by increasing the amount of water in the river, which is not easily done. Furthermore, river meanders actually migrate *downstream*; by erosion occurring on the downstream sections of the meander, just past the outer arcs, as can be seen in fig. 5-14, labelled "bank erosion," past the outer arc, and obviously eroding downstream of it.

Despite the fact that T. J. Randle shows in his diagram (fig. 5-14) that erosion is on the downstream section of a meander bend, and that meander amplitude is always constrained by river channel width, which is dictated by water volume, i.e., by the laws of physics, his paper opens with the surprising statement: "The migration of river channels across their floodplains . . . is a natural process."

No, it is not; it is an imaginary one based on a requirement of Charles Lyell's theory of river-valley formation, and is simply dogma accepted as true. Blindly accepted, I might add, since T. J. Randle and many like him seemingly cannot see that their own work contradicts it, as is evidenced by the only conclusion one can come to from fig. 5-14, provided by Randle himself, plus the known physical limitations on river geometry.

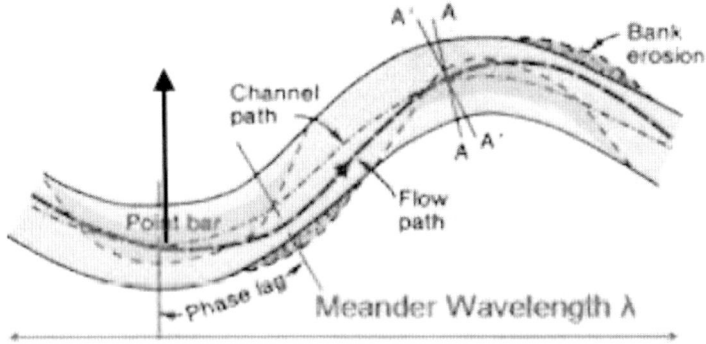

Fig. 5-14: Schematic meander showing Planform Phase Lag, erosion areas and point bars (deposition). The greater the phase lag, the greater the down-stream migration of the meander. Erosion is not at the outer arc of the bend, but downstream of it, and hence, the meander migrates downstream. Note that no erosion is shown at the centers of the outer arcs, and this would be a necessity for the meander to migrate across the valley. Even so, almost all writers persist in referring to lateral migration via meander erosion, despite also knowing that the radius of a meander bend is limited to 2-3 times the channel width (arrow) which cannot change without an increase in water volume. This automatically limits the meander amplitude and the ability of the river to erode in a direction so as to widen or move the meander belt. Of course, as for any river, some event such as an irreversible levee breach may also cause the entire meander belt to alter course. Diagram: Randle, T. J., 2006, USGS.

There does come a point where the floodplain becomes so flat and extensive from thousands of years of deposition that the meander system of a river loses all its integrity. In this case the river meanders, or more accurately, just wanders all over the place, often flowing around in loops to almost rejoin with itself. Such rivers are seen especially in Siberia, and the rivers flowing through the Pripet Marshes of Belarus/Ukraine, which is shown later in chapter 8, seem headed to the same condition.

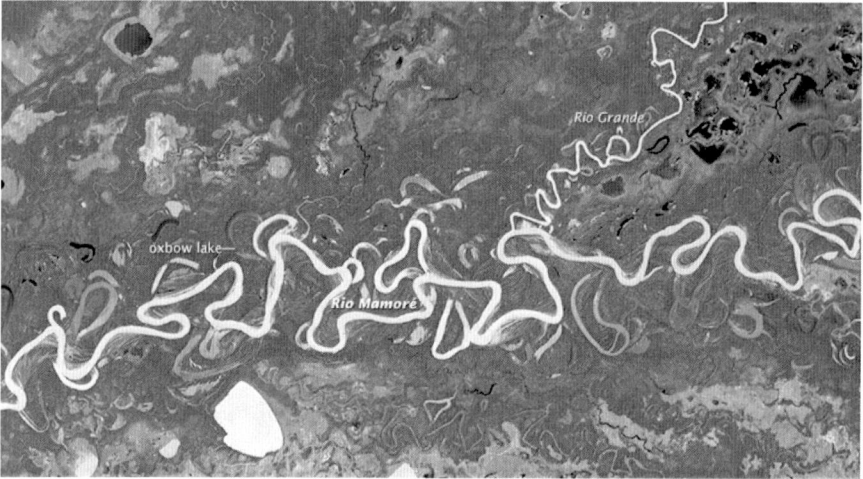

Fig. 5-15: Meander Belt of the Amazon River and the tributary Rio Grande. As can be seen from the scars of old meanders, the river has kept to the same general course. Image: NASA Earth Observatory.

The final word on the issue goes to Albert Einstein, an authority on physics if ever there was one. In 1926 he published a paper, "The Cause of the Formation of Meanders in the Courses of Rivers and of the So-Called Baer's Law," in *Die Naturwissenschaften*. This paper was modified and republished by Kent A. Bowker of Chevron, U.S.A., Inc.

In the paper, Einstein describes the mode of flow of water (fluid dynamics) in a river and identifies the ways in which the water moves at different speeds at different levels and at different places in the typically sinuous course of the river. The velocity is higher at the top of the flow than at the bottom, due to drag of the river bed, and hence a rotational, or helical, flow pattern is set up. In a meandering river, because the higher velocity portions of the stream will be driven to the outside, concave, portion of the river bend, erosion will be greater there.

But, because helical flow possesses inertia, the circulation (and the erosion) will be at their maximum *beyond the inflexion of the curve* (shown as Bank Erosion in fig. 5-14). Hence, the wave form of the river will *migrate in a down-current direction*. Finally, Einstein explained that the larger the cross-sectional area of a river channel, the slower the helical flow will be absorbed by friction, which explains why larger rivers have meander patterns with longer wavelengths (Bowker, 1988) and hence, longer bend radii.

This work was repeated by other scientists, notably Leopold and Wolman (1960) who confirmed the correctness of Einstein's analysis and whom Randle himself (fig. 5-14) referred to in his paper, the 1964 edition of Leopold and Wolman's well-known and widely-used book, *Fluvial Processes in Geomorphology*, which I also consulted.

Leopold & Wollman more or less repeat what Einstein said about meanders, particularly about water velocity being highest downstream of the outer arc, in that: "The highest velocity at any point tends to be located near the concave bank just downstream from the axis of the bend" (page 299). They go on: ". . . where bank erosion is most frequent is just downstream from the axis of the bend" (p. 301) and finally: ". . . river curves tend to move downvalley" (p. 301) (Leopold & Wollman, 1964). Therefore, these writers are clearly confirming more or less everything important Einstein said about river meanders, leaving us to wonder, somewhat mystified, at T. J. Randle's claims.

Further, while a meander bend migrates downvalley, so too does the point bar inside each bend. Material eroded from the outside bank is deposited on the downstream side of the point bar, while the upstream side of the point bar is eroded per the phase lag, as shown in fig. 5-14. Hence, the point bar also migrates downstream, keeping within both the meander bend and the meander belt.

However, it would seem this transfer of sediment from one bank to the other, which only occurs in a meandering river, is the basis for the notion that rivers can migrate across their valleys and is likely the reason Lewin et al. (1991) proposed the "sediment-laden, fast-moving, meandering river" in the first place, even though such a thing is patently impossible.

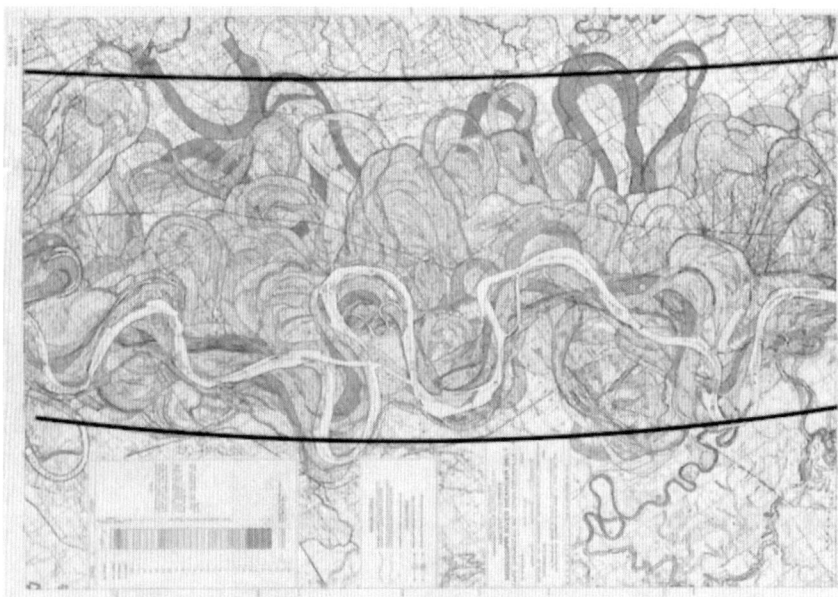

Fig. 5-16: Ancient Courses (of the) Mississippi River Meander Belt by Harold Fisk, 1944, for the U.S. Army Corps of Engineers. Famous map of the many courses the Mississippi has followed over the last few thousands of years, and similar to the Amazon. As can be seen, the river has changed course any number of times, but has always remained within a clearly defined meander belt, roughly defined here in black. Despite the obviously flat land, showing fields on either side, there has been no "lateral migration across the floodplain," so we can take it for granted that the Mississippi does not migrate across its own valley, and did not cut out that valley either—it's been filling it, ever since it started flowing. Image: H. Fisk, U.S. Army Corps of Engineers.

In the fairyland that is geology, however, it seems any river can do whatever the geologist wishes it to do, and in the Konitsa Plain, the Voidomatis River of the long, long, ago and the far, far away, could magically and continuously change its course sideways. According to the authors, the river "aggraded," or built out sideways the coarse gravels lower down in the channel, and the fine sands and silts on top.

Despite the fact that the banks of the river are 10 meters high, the idea here is that the river would, or could, for some reason or other, continuously eat away at one bank while depositing against the other bank, as it shifted its course laterally across the valley, very contrary to Einstein's definition of the mechanics and dynamics of the meandering process, and of course, never seen in action.

This concept is also very contrary to the transporting mechanics of a river; it must flow rapidly to carry heavy gravel sediment, but cannot meander unless it's travelling slowly, in which case it cannot carry heavy gravel sediment, and it certainly can't wash away coarse gravels. Nor, mind you, can it flow uphill.

Lying behind this so-called interpretation is a serious historical distortion and misrepresentation of the process and mechanics of river meandering, first promoted by Lyell himself, who claimed that all rivers migrate back and forth across their valleys, and in so doing actually cut out the valleys. This, again, was a very much debated concept in the 19th century and many rejected the idea, especially the older geologists.

However, today, this is a major part of the standard explanation for river-valley formation, fed to every schoolchild and college student for the last 150 years or so. No doubt Woodward et al., and the rest (including myself) were also taught this, and like most geologists, and everyone else, they still believe it. That I don't is not to say that

I am somehow specially gifted and possessed of unique powers of insight and interpretative talents; it's just that I was given cause to doubt, and I did, as I described in the introduction.

In any case, the authors propose that this is precisely what that fast-flowing, sediment-laden Voidomatis river did in the Konitsa Basin, all the while flowing through the plain to join up with the Aoos river coming from the east. This explanation for coarse gravels covering a wide plain originated back in the 1830s, in efforts to explain, without the use of a "deluge," the same, or similar, coarse gravels covering the valley through which the sluggish river Thames now flows, a river presently incapable of moving any of the coarse, cobbly gravel material of its bed, something it has in common with many rivers.

To account for all this sedimentation of large deposits of coarse material, extending well down the valley and out onto and across the Konitsa Plain, all authors must invoke floods. Woodward et al. (2008) give something of a synthesis, mentioning floods in connection with sediment delivery, i.e., *major meltwater events*, *large floods* and such phenomena as *largest meltwater floods* and *frequent large floods*.

These phrases are often used in conjunction with *enhanced sediment contributions*, which is to say the river washed away or picked up much more sediment. The point here is that extra-large floods are the only agents that the authors can come up with to explain the evidence from the Voidomatis, because they have no choice.

What is notable is that all these workers found only an identifiable total of three major flood deposits, the latter two dating to the post Ice Age period. In general, given that Greece didn't have an ice sheet, the annual spring thaw in such a glaciated region should produce a single identifiably individual layer for each yearly flood, as we see at river deltas and mountain valleys today after the Spring thaw, and it should be no different in the Voidomatis Valley, where, according to Woodward et al. (2008) deposition has been going on for over 150,000 years.

Contrary to Lyell's uniformitarianism, no evidence exists, or has been reported, that in any way indicates that any long-term, continuous deposition has been going on in the Voidomatis basin; all that is mentioned are a few flood "events" and a few major units.

Thus, in the Pindus Mountains, the evidence has been interpreted as indicating three phases of glaciation (plus one, an outlier, Hughes et al., 2006). The same workers have also identified three phases of (flood?) deposition, one during the Ice Age and two following it, which is to say over the course of maybe about 1 million years, though they offer no evidence for this time period. And we must not forget that in the Aegean, the deposits in the basins were often found to comprise three or four units below the recent normal marine sediments.

It is a very curious coincidence indeed between sensible official, and conventional geological science and the crackpottery of that old Egyptian priest, and not to mention Plato and the rest of the ancient Greeks, who all firmly believed that their country had suffered three or four floods. Also, I would guess that the flood dated here to *within* the Ice Age actually occurred at its close, and we know when the more recent one occurred.

Since we treated the Pindus Mountains in detail because they are the best studied, and the most glaciated, we will now take a relatively brief overview of other valleys and plains in Greece, and compare the evidence. In general, other river valleys don't differ all that much from what we've seen already, but there are certain phenomena well worth examining.

Chapter References

Bailey, G. N. et al., 1990, The "Older Fill" of the Voidomatis Valley, North-west Greece and Its Relationship to the Palaeolithic Archaeology and Glacial History of the Region, *Journal of Archaeological Research*, Vol. 7, No. 2, pp 145–150.

Bishop, P., 1995, Drainage rearrangement by river capture, beheading and diversion, *Progress in Physical Geography: Earth and Environment*, Vol. 19, Issue 4, pp 449-473.

Bowker, Kent A., (Chevron, USA, Inc.), 1999, Albert Einstein and Meandering Rivers, *Search and Discovery* Article #70001.

Davis, W. M., 1899, The Geographical Cycle, *The Geographical Journal*, Vol. 14, No. 5, pp 481–504.

Einstein, Albert, 1926, The cause of the Formation of Meanders in the Courses of Rivers and of the So-Called Baer's Law, *Die Naturwissenschaften*, Vol. 14, pp 105-108.

Fan, N. et al., 2018, Abrupt drainage basin reorganization following a Pleistocene river capture, *Nature Communications*, Vol. 9, 3756.

Hughes, P. D. et al., 2006, The glacial history of the Pindus Mountains, Greece, *Journal of Geology*, Vol. 114, pp 413-434.

Lauder, W. R., 1968, Stream Capture, Piracy. In: *Geomorphology. Encyclopedia of Earth Science*, Springer, Berlin, Heidelberg, pp 1054-1057.

Leopold, L. B., Wolman, M. G. & Miller, J. P., 1964, *Fluvial Processes in Geomorphology*, W. H. Freeman and Company, San Francisco, CA.

Lewin, J. et al., 1991, Late Quaternary Fluvial Sedimentation in the Voidomatis Basin, Epirus, Northwest Greece, *Quaternary Research*, Vol. 35, (pp 103-115).

Lyell, C., 1832, *Principles of Geology,* London, John Murray, Albemarle St., Ch. 2.

Randle, T. J., 2006, Channel migration model for meandering Rivers, *Proceedings of the Eighth Federal Interagency Sedimentation Conference* (8th FISC) April 2-6, 2006, Reno, Nevada, USA.

Shugar, D. H. et al., 2017, River piracy and drainage basin reorganization led by climate-driven glacier retreat, *Nature Geoscience*, Vol. 10, 370-375.

Stokes, M. F. et al., 2018, Ongoing River Capture in the Amazon, *Geophysical Research Letters*, Vol. 45, Iss. 11.

Tourloukis, V. & Karkanas, P., 2012, Geoarchaeology in Greece: A Review, *Journal of the Virtual Explorer*, Vol. 42, No. 4.

Willet, S. D. et al., 2014, Dynamic reorganization of River Basins, *Science*, Vol. 343, Issue 6175.

Woodward, J. C. et al., 1992, Alluvial sediment sources in a glaciated catchment: the Voidomatis basin, northwest Greece, *Earth Surface Processes and Landforms*, Vol. 17, 205-216.

Woodward J. C. et al., 2008, Glacial activity and catchment dynamics in northwest Greece: Long-term river behavior and the slackwater sediment record for the last glacial to interglacial transition, *Geomorphology*, Vol. 101, 44-67.

CHAPTER SIX:

RIVER GEOMORPHOLOGY AND RIVER BASINS

IN THIS CHAPTER, WE CONTINUE WITH THE RIVERS of Greece, following on from the Voidomatis that we examined in some detail. Here, we will take something of a tour of various rivers and valleys in different parts of Greece, from the Peloponnese in the south to the mountains in the north, without going into great detail on any one of them. We will find, however, that practically all of them have associated features and phenomena that are rather difficult to account for, and we'll come across some interesting "uniformitarian" explanations.

We saw the Voidomatis, with its terraces at extreme heights relative to the present river, the foreign material (ophiolites) in its valley, and the inexplicable, coarse clastic deposits filling the flat plain beyond the mountains, a plain that, had it any normal, modern-day river, should only contain fine alluvial material. Also, we saw very large and rounded boulders high up in the headwaters of the river which are rather difficult to explain. It will be of little surprise when we find that other rivers have similar puzzles for us, along with different ones all their own.

As it was on Mt. Tymphi, Smith et al. (2006) report that Mt. Olympus to the east, also displays three discrete "stages" of deposition, related to the same three glacial stages. Glacial sediments consisting of tills are found on the heights, while fluvio-glacial and simply fluvial sediments are found on the lower slopes and nearby lowlands (piedmont). Diamictons are mentioned along with "bouldery" till, and what the writers call moraines, consisting of a basal till overlain by an: ". . . upward-fining sequence of (glaci)fluvial sediments," the latter sounding very much like a typical flood deposit.

The writers also mention earlier workers as having identified ". . . three discrete sedimentary units (units 1-3) that record deposition during the Quaternary. These three units have been considered alluvial fan deposits . . . without any direct glacial influence." Again, we find three sedimentary units, but, curiously, no "three glacial phases" here.

Of interest also is a mention that "glacial sediments of the Xerolaki-Mavroneri River Valley can be traced continuously from cirques on the north face of Mt. Olympus to the vicinity of Katerini on the Aegean coastal plain." Here then, we find much the same situation prevailing as in the Konitsa Plain of the Voidomatis, and the writers also tell us that such sediments merge with others that came from the High Pieria Mountains to the north.

This latter source is confirmed by K. Tsanakas et al. (2019) in their paper on the geomorphology of the Pieria Mountains. The study area is shown outlined by a square in fig. 6-1, a hill-shaded map of the area. Amusingly, while displaying little geological information on this map, they do show two (relatively) major rivers and clearly indicate river diversion in the case of the Aliakmonas, with an arrow pointing to the Meteora Conglomerate, which is a famous and rather strange rock formation, supposedly formed 23 my ago.

The processes being described here in the Pieria mountains by these writers take the usual millions of years, as we can expect. In any case, the writers are implying river diversion to the northeast. They consider the diversion as due to "uplift" at some point in the past, which they do not specifically describe, but which is very much in keeping with the Davis cycle. The Meteora conglomerates, however, do indicate a large flood in the unknown distant past.

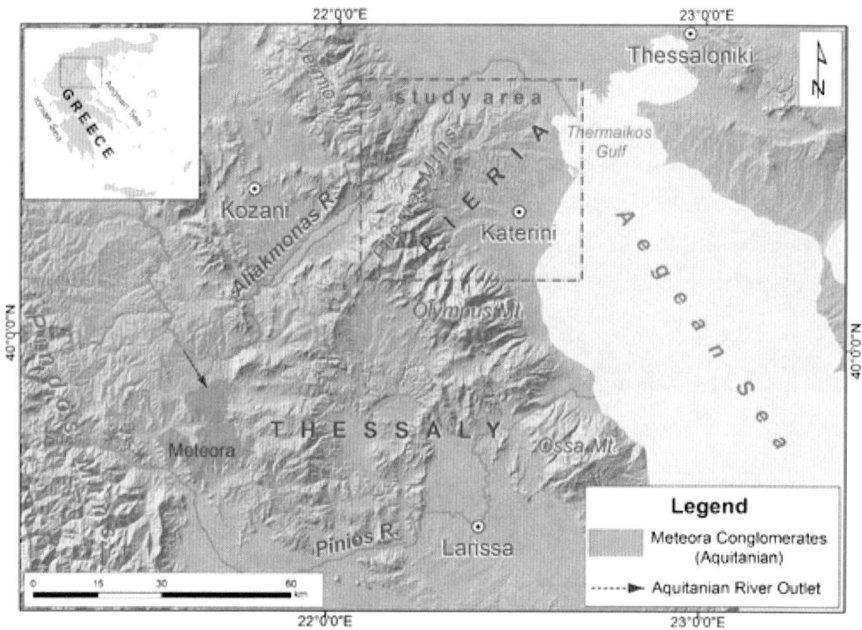

Fig. 6-1: Hill-shade map of study area showing the relationship of Pieria Mts., Mt. Olympus, and the Aegean Sea and also the coastal plain near Katerini. Image: K. Tsanakas et al., 2006.

In the Petriotikos stream valley are found, again, "fluvio-torrential" deposits in the form of extensive terraces along the main channel, where these writers also report a "stream flow shift" due to a recent activation of a fault, i.e., tectonic movement, which would explain such diversion quite uncontroversially, if true. Alluvial fans, reported from the north side of Mt. Olympus, are composed of fluvial sediments containing medium-sized, well-rounded pebbles, quite in keeping with the term *fluvio-torrential,* which vague term itself remains unexplained, though it clearly implies flooding of some severe kind.

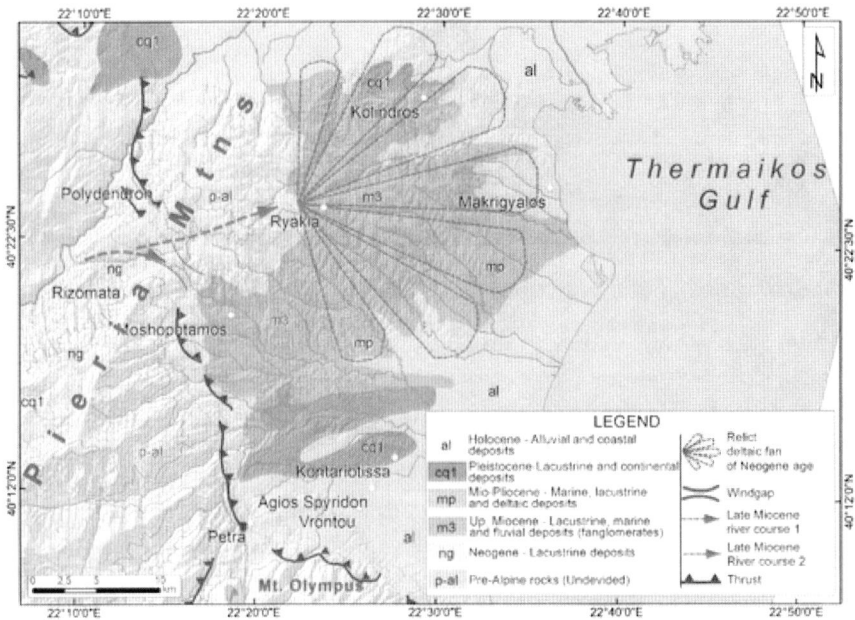

Fig. 6-2: Lithological map of study area. Former Aliakmonas River shown by dashed lines, dated as Late Miocene (5-6 my) by the writers. Needless to say, I don't accept all these millions of years; all I see in the map above is evidence for "major flood events." Image; Tsanakas et al, 2019, p. 500.

The most interesting findings and proposed explanation pertain to the top ridge and eastern side of the Pieria Mts., shown on the map in fig. 6-2. The areas on the map, labelled m3, and over-drawn in the east by the fan-shaped outlines, represent what the writers have determined to be alluvial fans. In fact, these are extremely large alluvial fans, with an arc perimeter of approximately 50 km (30 miles) and 20 km (12 miles) respectively.

The writers state that the ". . . composition and size of this (larger) fluvio-torrential fan points to a large river with high discharge rates and sediment yields." This former river, entirely theoretical, is shown by the upper dashed line, on the map, crossing over the ridge of the Pieria. The line below it is an even ". . . older channel outlet located at a higher elevation than the large deltaic fan," by which he means the smaller m3 unit to the south-west of the larger. They identify this former river by the presence of a "wind-gap," which is conventionally regarded as having been cut by a river, here the former Aliakmonas river cutting out its valley, and a typical wind-gap does look like it was formed by water. The rather puzzling thing about them is that they're often found on ridges, as though a river flowed up and over said ridge.

We can see by figs. 6-1 and 6-2, therefore, and from the paper itself, that Tsanakas et al. are suggesting that some former Aliakmonas river first flowed in a more southerly direction toward the Meteora conglomerate (fig. 6-1). Then it changed course, to the east-south-east, to flow over the top of the Pieria Mts. to deposit the smaller of the alluvial fans.

Then it changed course again to the east-north-east, to deposit the larger of the alluvial fans, here rather coarse and implying a powerful river, or flood, given that the writers refer to them as "fan-glomerates," before somehow changing its direction yet again, to the northeast, to excavate out an entirely different and very large valley for itself north of the Pieria Mts. and more or less at right angles to its original course.

That is a lot of course correction and excavating, not to mention transportation. Tsanakas et al. explain this physically impossible sequence of events as all due to tectonic uplift, à la Davis. And while I would not dispute tectonic activity, the "uplift" is questionable, and it does not automatically lead to erosion or river diversion.

Apart from these former wandering rivers, the authors also describe some of the other rivers draining the Pieria Mts. and Mt. Olympus, one such being the Xirolaki stream, which flows from the north side of Mt. Olympus and also displays evidence of torrential flooding in the form of alluvial fans, as shown in fig. 6-3.

Fig. 6-3: Xirolaki stream alluvial fan deposit consisting of semi-consolidated, medium-sized, well-rounded pebbles intercalated with fine-grained deposits. Image: Tsanakas et al. (2019).

If we examine it, we notice that each of the layers lower down consists of a coarse layer with gravel and pebbles with a fine layer above it, and with some (narrow zone) of grading between them. As mentioned, Tsanakas et al. identify the unit as an alluvial fan deposit, the layers simply being flood sequences. The thick top layer is comprised almost entirely of pebbles and cobbles with some bigger stones in what appears to be a matrix of sand or silt. While some red discoloration can be seen at the bottom of it, this upper unit looks like one massive layer, clearly flood deposited.

While the evidence of the alluvial fans and fanglomerates relating to the Pieria Mts. and the Xirolaki stream certainly implies large amounts of water acting in the past, no grounds have been given for making the case that some "earlier version" of the Aliakmonas could have been any bigger than the present one. So, what exactly was this "large river with high discharge rates and sediment supply" that Tsanakas et al. have drafted in to do the work? They tell us there was very little ice on the Pieria Mts. during the Ice Age, but presumably there was snow, and so spring meltwater floods could not have been large, no larger than any flowing from any Alpine glacier today.

Those floods do not bring boulders and cobbles down the Alpine valleys at present. Also, rainfall could hardly be expected to be all that greater than today either, given that moisture evaporation from the Mediterranean during a cooler Ice Age would presumably and logically be reduced. Vague references to ill-defined tectonic uplift are of little use, and the writers have no reasonable source for all the required water. However, something obviously supplied it.

As can be seen in fig. 6-1, the Pinios River flows from the Pindus Mts. to the south of Mt. Olympus and across the Thessaly Plain, or basin, to enter the Aegean just to the north of Mt. Ossa. Migiros et al. (2011) give a general overview of the geomorphology of the river, noting a number of terraces in places along the valley, typically no more than three, and usually only one or two, with the lowest one being due to recent and ongoing normal flooding. We met terraces before in the Voidomatis valley, shown in fig. 5-13, and we will discuss them further in a later section of this chapter.

The highest terrace, composed of mud, sands, and gravels, is only 8 m (25 ft.) above the river, and is located well down the river at the Larissa plain, where, in fact, it forms a fairly extensive floodplain, up to 1,600 m (1 mile) wide. This would imply that the river, or whatever body of water deposited the material, was much deeper than 8 m, and a good deal more powerful than today to lift gravel onto a terrace.

This terrace has been determined to date from Ice Age times, while the later two are dated as Holocene and even more recent (Migiros et al., 2011). The floods experienced today in the Pinios typically do not, at their worst, rise much above the 8 m level of the upper terrace, and only areas near the river are flooded, meaning, as usual, that a lot more water must have flowed here at some point in the past.

The Larissa Plain is, like most other plains in Greece, a filled-up tectonic basin, the usual graben. And while the sediments in the river itself, as shown in fig. 6-4, comprise the lower layers of valley infill, the plain itself, according to Caputo et al. (1994) is up to 10 m (30 ft.) above the river, and it, and the whole drainage system, are considered to have "evolved" by the repeated down-cutting of this channel into those stony sediments, due to tectonic uplift. However, fig. 6-4 shows no evidence of any down-cutting into the moss and algae covered stones of the riverbed.

Fig. 6-4: Pinios River in the valley of Tempi. The moss on the stones in the foreground suggest that the river neither moves the stones nor erodes them. Note that all the stones visible appear to be of a similar size, indicating that the size-sorting was done by a much larger river than today's little trickle. Image: ID 98259447 © Vprs26, Dreamstime.com. Royalty-free Lic.

This process is known as "rejuvenation" and occurs after a river has reached a certain stage of "maturity" and is then affected by a "tectonic movement" that causes the course to become steeper, and the river to flow faster again like it did when it was a youngster.

These terms, youth, maturity, evolve and rejuvenation, come to us compliments of the same W. M. Davis we met earlier. As virtually everyone who has commented on his "Geographical Cycle" says, Davis was greatly inspired by the Darwinism that was sweeping Britain and America in the late 19th century. For some unfathomable reason he applied its concepts to landscape "evolution," along with Charles Lyell's concepts of the different stages of existence (lifecycle) of an entire species (youth, maturity and old age). See for example (Flemal, R. C., 1971).

While the use of all these familiar terms might sound appealing and might appear to be sound and appropriate analogies, they are so only because these terms have been sold to academia and the public as being all very scientific. "Sold" is probably an understatement; "drilled in" by repetition and indoctrination is more like it.

The trouble is that, despite widespread *belief* in their veracity, neither erosion by any river, nor downcutting (into solid rock, especially) nor Darwinian evolution, nor Lyellian concepts of species lifecycles are anywhere seen to be occurring at present. Nor can it be proved, from the evidence, that they have occurred in the way orthodox geology claims, anywhere, ever.

Species have lived and died out, yes; animals and plants have changed form over time, yes; river channels have been cut out and deepened, yes, but none of these can be demonstrated to be *in action*, or from any evidence, to *be occurring* or to *have occurred* in any kind of slow "evolutionary" way. Therefore, an unconscious and preprogrammed "leap of faith" is always required for their acceptance, and such acceptance, in the absence of evidence, is merely a function of academic and media-driven indoctrination and the resultant habituation.

Darwinism, of course, is assumed, by now and by most, to be essentially received wisdom, beyond question, having been proved out long ago, and only crackpot creationists don't accept it because they're blinded by religious belief, and so on and so forth. However, neither the public, nor the students in our schools or universities are ever told just how far away from proved any of it actually is.

Caputo et al. (1994) do not tell us much about the sediments in the plain but do mention two alluvial deposits of Holocene age (post Ice Age). These "units" clearly parallel those we heard about from van Andel et al. (1991) earlier. Caputo refers to these authors and the dates of 14,000 to 10,000 years ago for the earlier "event" and about 7,000 to 6,000 years ago for the later one. The latter is a bit older than the ages (bracketing 5,000 years) obtained by van Andel et al. (1991) and Zangger (1992), who put the later "alluviation event" at approximately 5,000 years ago, the date that keeps on appearing. In any case, dating or age differences of one or two thousand years mean little enough, as we'll see, and despite all the claims.

Note that contrary to the very conservative behavior of a river along its course inland, as we discussed earlier, and its resistance to lateral migration, at the coast, the situation is very different. Due to sediment deposition going on where the river meets the sea, and interference from marine action, which is to say "outside interference," channels may change direction fairly often due to the changing geometry of the delta, i.e., as they are forced to, as per Clausewitz. The Mississippi delta is an excellent example of this, as contrasted with its behavior upriver.

Moving to the western coast of Greece, to the Patras (Patraikos) Gulf area in the northwestern Peloponnese, we have two papers on the geomorphology and deposits of the coastal plains on the southern side of the gulf, as shown in fig. 6-5. As can be seen, the plains are rather small and bounded by hills, from which there is a fairly abrupt transition from the slopes to the flat land of the plains.

Taking the left (north-facing) plain first, Alevizos and Stamatopoulos (2016) describe fluvial and alluvial deposits of considerable thickness and variability, which they then "explain" as relating to climate changes and their effects, without actually specifying those changes, or their effects.

The supply of debris is explained by frost-shattering during the Ice Age, mass-movements, and subsequent sediment over-charging of the streams, leading to the build-up of sediment in the plain. Uplift, again, is invoked to supply the necessary velocity and transporting power to the rivers, but "flash-flooding" and "extreme flash-flooding events" are again necessarily called upon to get over the awkward fact that no present-day flood, or even inferred Ice Age meltwater flood, could have filled-in this basin and spread coarse "fluvial" and "alluvial" sediments all over it to form the plain. As can be seen in fig. 6-5, what rivers are there are small and flow in shallow valleys.

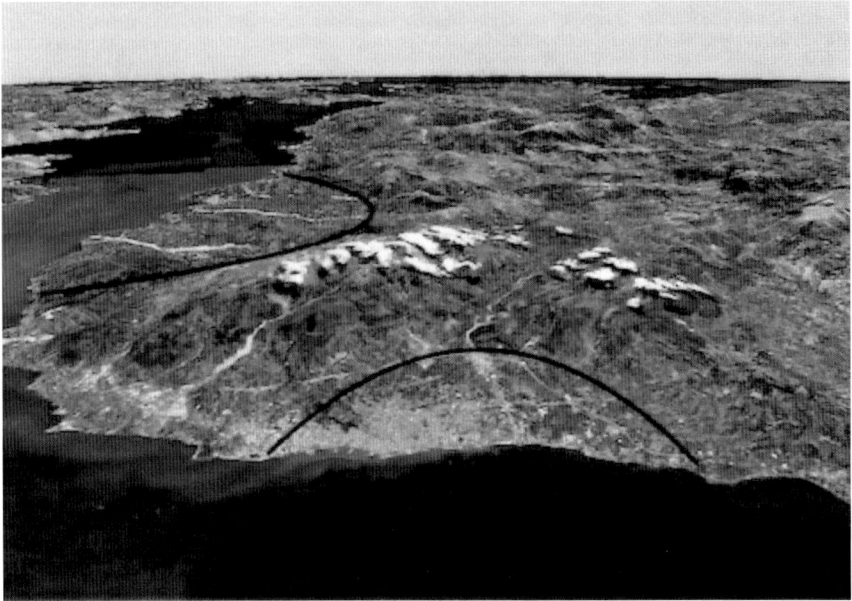

Fig. 6-5: Coastal plains in the Patraikos Gulf area. The plains are sharply bounded by hills with only a few small streams flowing into and through the plains, making the source and supply of all the sediment questionable. The little valleys behind these plains could not supply much rain or sediment. Image: Patras NASA Worldwind, CC BY SA 3.0.

Hence, they have limited catchment basins and low gradient, implying limited water volume and sluggish flow, which could not have been much different, if at all, in the past. Thus, the question is where is all the water to come from to produce these "extreme flash-flooding events"?

One river flowing through this plain is the Peiros. Alevizos et al. (2018) give a brief description of its geomorphology. The area has apparently suffered subsidence, followed by slight uplift. However, they tell us the coastal sands overlie lacustrine and marine mud, so it must have been more than slight uplift, since sea level supposedly rose some 125 m (400 ft.) since the Ice Age, so this marine mud must have been raised up a good bit more than that to be exposed on the seashore.

The mention of "lacustrine," which means *freshwater* lake, is interesting considering we're on the seashore here, so how could there have been a freshwater lake and mud? We'll get to an interesting interpretation later from geologists studying mega-flooding in the Mediterranean, and from a 19th century Englishman, who wasn't as severely infected with uniformitarianism as many of his compatriots.

The writers, Alevizos et al. (2018), mention the usual fluvial sediments and, also, alluvial fan deposits (floods). The mud and alluvial fans are "unconformably" overlain by terrestrial fluvial deposits, meaning, as usual, that erosion occurred between the two. These fluvial deposits consist of: ". . . alternating layers of conglomerates, sands and silt fining upwards," which, as we now know, is the classic multi-flood sequence, in this case floods over alluvial fans as seen earlier in fig. 6-3. The presence of basal conglomerates imply rapid water movement, as does the erosion of the underlying material to produce the unconformity forming the base beneath those conglomerates.

The other plain, in the lower foreground above, is the Patras City coastal plain. Rozos et al. (2006) prepared an engineering geological map of the area, along with an assessment of the sediments of the plain, to aid in foundation design. Thus, they focus only on the sediments. The first thing they tell us is that the Quaternary sediments (Pleistocene and Holocene) are both alluvial and *diluvial* and very heterogenous. And so, here we have another use of the word "diluvial" by a modern-day geologist, considering it normally refers to floods of the type described by the supposedly bible-thumping geologists of the early 19th century, and as explained by Ball and Buckland in the Epilogue to volume 1.

The sediments are generally the same as elsewhere: fluvial sands and gravels, fan deposits containing gravels and cobbles, conglomerates and fine sands, silts, and muds. These sediments are up to 80 m (270 ft.) thick, and the fan deposits are very coarse-grained with little fine material. The cobbles in these fans are reported as being from 15–20 cm (6–8 ins.) in diameter, which is a fairly large size, and this means those fans are without doubt flood deposits, and not from any kind of ordinary flood, either, as we know by now.

They do give a brief note on the Patras graben that forms the basin of the plain, telling us that it is infilled with 1,000–2,000 m (3,300–6,600 ft.) of terrestrial, freshwater sediments (no more precise description) and a further 1,200 m (4,000 ft.) of ". . . fan deltas and coarse-grained terrestrial sediments . . ." Three thousand meters (9,000 ft.) is a lot of sediment, and freshwater sediments next to the sea are curious.

At Mt. Chelmos, where Woodward & Hughes et al. studied the Glacial history, Pope & Hughes (2014) followed up by studying the connection between glacial melting and sedimentation, as they and others did for the Pindus Mts., via the Voidomatis river.

Here at Mt. Chelmos, they report the same three moraines in the highest regions, the last one dating to 12 to 11.5 thousand years ago, the critical period. They also report valley moraines and extensive glacio-fluvial fan systems, terraces, and glacio-fluvial outwash deposits.

Thus, the glaciation of Mt. Chelmos has, like it did for the Pindus Mts., produced large volumes of debris. While they date the last phase of glacial melting to about 11,500 years ago, they date the gravels in the valleys to only 6,000 to 7,000 years ago (5,000?) meaning the ice didn't melt for 5,000 years, which makes no sense—especially since Mt. Chelmos is well to the south in the Peloponnese, which is a bit far south for ice in the first place, even in an Ice Age.

Stamatopoulos & Alevizos (2018), whom we met in the Patraikos Gulf, also worked in the Mt. Chelmos area close to where Pope & Hughes worked, namely the Kalavrita Plain, specifically the Vouraikos River. In a paper on the geomorphological evolution of the river system, they tell us that the plain is an infilled graben, the area was tectonically active recently, terraces are present in the valley, and the sediments reported are generally much the same as everywhere else.

Cyclic flooding events are reported, and archaeological sites were uncovered. The deposits show very poor sorting and particle sizes range from clay to gravel and even boulders up to 2.5 m (8 ft.) in diameter, making these deposits sound like extremely coarse diamictons, which, being down in the valley, cannot be "glacial tills" but must be flood deposits. Again, these sound very much like Bald and Buckland's "old alluvium" or "diluvium."

The authors, however, choose to suggest "debris flows" as their cause, without being considerate enough as to say why, exactly. The clasts in this "debris flow" are fairly well-rounded which suggests they were transported by water, and not simply a water-logged landslide, as they would then logically be much more angular.

It is, I suppose, possible that it was a very well-traveled debris flow, but as we saw earlier, such deposits can be difficult to identify, and it would indeed need to be a *very well-travelled* debris flow to actually round off a large boulder. At the same time, the authors note "... more than one incidence of debris flow ..." just like we had more than one incidence of "glacial till" and "extreme flood" and "diamicton," and so on with all the rest. As we saw in chapter 4, a debris flow is little different in final appearance to the bottom-slurry dragged along by a major flood, and, considering everything else we've seen, I'm inclined to think that that's what we're dealing with here—major floods.

The authors have some trouble explaining other phenomena found in this valley, not the least of which are the terraces, also found in other valleys we examined. Terraces are found on one or both sides of a river and have their upper surfaces lying at an elevation well above that of the river itself, much higher than the typical floodplain. The terraces here, like the others we have encountered in Greece, are all very high up, and these have all proved somewhat more difficult to explain.

They must, by uniformitarian imperative, be explained only by means of a river flowing in the valley, whether or not it was the river and the valley in their current form or some previous and different forms, though, of course, a different form for this river and valley in the past would seem to violate one of the first principles of uniformitarianism.

The crux of the matter is that, conventionally, some version of the river that is flowing there *at present,* must first have deposited the material by filling the valley up to the level of the topmost terrace, and then that same river, due to some kind of change having occurred, always involving "tectonic uplift" which made it flow faster and stronger, must have come along and eroded, or "incised" down into the material to make a new deep central channel, and leave the terraces at the valley sides, and this process of uplift and downcutting is repeated any number of times to produce the stepped arrangement as shown in fig. 6-6.

The trouble is that typical terraces do not show such a geometry, or "morphology" as it's called. What the standard model shows is a single mass, progressively eroded into a step-like form, as shown in fig. 6-6. However, in reality, the highest terrace is typically *stacked against* the valley sides and lower terraces are each *stacked against* the next higher one, as shown in fig. 5-13 and hence, terraces must also have a different origin, which they do.

In the other model, which shows terraces stacked against each other, as in fig. 5-13, the river first fills in the valley in the usual uniformitarian way with the river "meandering" back and forth across the valley, as per Lyell. Then, after the usual tectonic uplift, the river comes along and removes the central section of fill, to leave two terraces against the valley sides.

Then another version of the river comes along and partially fills up the central area. Then comes tectonic uplift and some other river comes along and again removes the central section of this smaller infill, to leave two lower terraces, and this process may be repeated to leave a third smaller pair of terraces, and so on, to produce the stacked terrace arrangement as seen in the Voidomatis Valley, according to academic orthodoxy's highly unlikely scenario. We'll be meeting other highly convenient up-and-down tectonic movements again later.

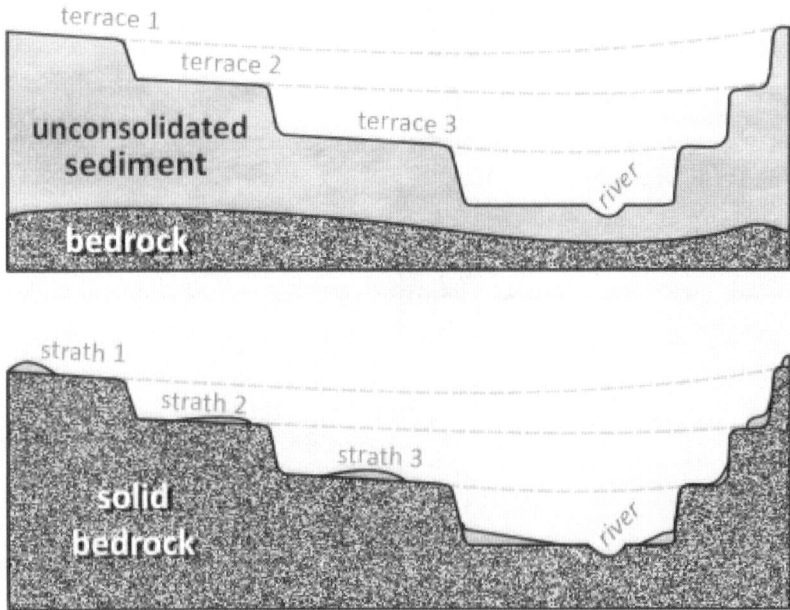

Fig. 6-6: Comparison of fluvial and strath terraces in a river valley. In the upper diagram, the little river supposedly once flowed at the elevation of the top dashed line, having deposited all the unconsolidated sediment to fill in the valley. Subsequently, due to three supposed uplifts to get the river flowing faster, it eroded away a lot of it in three distinct steps. In the lower diagram the river simply eroded away all that solid rock, again in three steps due to supposed uplift, leaving remnants of sediment on the strath terraces. Image: American Geophysical Union.

We will leave the subject of terraces for the present, but we must mention the case of the "strath terrace," which is more of an immediate problem. Alluvial or fluvial terraces are entirely composed of sedimentary deposits only, while strath terraces are flat surfaces eroded into the underlying bedrock and often having a thin coating of fluvial or alluvial deposits, as shown in the comparison diagram in fig. 6-6.

The standard, so-called explanation for both types of terrace formation is successive periods of uplift, after a period of deposition in the case of fluvial terraces and a period of bedrock erosion in the case of strath terraces, the bedrock having first been eroded to some form of "peneplain" as per Davis. Uplift then caused the river to flow faster and erode instead of depositing, and, in the case of the above schematic, this occurred a further two times, to produce the profile seen here.

This, again, is the theoretical explanation, as proposed by Davis, and many others before and after him, despite a lack of any external evidence confirming said uplift (external evidence being evidence from anywhere *outside* of the valley in question). This problem caused no end of debate for decades in the 19th century, a debate which spawned any number of fantastic theories and imaginary processes and agents.

In fact, geologists have now gone considerably further in their interpretation of terraces, straths, and sequences of same. Given that terraces are now *assumed* to be formed by uplift and re-erosion, terraces themselves are now used as proof positive that uplift has actually occurred, and that by "estimating" the rate of erosion, "estimated rates" of uplift can be obtained (see Litchfield & Berryman, 2006), providing us with yet another excellent example of the circular reasoning that we met with in volume 1 with the marauding Indians.

To get back to the Vouraikos river, Stamatopolous & Alevizos (2018) declare their intent to "reconstruct the fluvial evolution of the Vouraikos river." They are tying it to the tectonic history of the area, the phases of uplift and incision as outlined above. There is little doubt about tectonic activity, so the writers claim that the activation of faults bounding extensional grabens (those WNW-ESE and NNE-SSW faults we discussed in the section on the Aegean) assisted the uplift-induced valley incision seen in the Vouraikos valley. Such incision into previously deposited fluvial sediments formed the fluvial terraces, as per standard theory, and this is simply presumed here, since no evidence is supplied.

The strath (rock) terraces, on the other hand: ". . . are the outcome of the interaction between the sum of the geomorphological processes that led to the bottom of the valley being carved by graded or steady-state longitudinal profile (meaning the river is at equilibrium, and simply flowing in some kind of ideal, or perfectly accommodating, bed) while the river is, somehow, synchronously lowering and widening its channel." They do not say how this lowering and widening occurs, while they are also telling us, in essence, that the strath terraces were formed by whatever it was that formed them, which is not very enlightening, I must say.

Thus, a major change must have come upon the river to cause it to incise deeper into its bed and leave the strath terraces high and dry. Such change typically and conventionally implies tectonic uplift, to put the river into disequilibrium, or climate change, to increase the water supply, or some other unspecified agent that induced some change.

Both of these agents and processes were invoked back in the 19th century to explain river and lake terraces, and both were ultimately thrown in the trash can, until resurrected by the likes of Davis, for the want of anything else acceptable to uniformitarianism.

The real key to fluvial and strath terrace formation is revealed by practically every writer being forced to invoke floods, megafloods, or deluges, etc., to explain the evidence found in the river valleys and plains of Greece, particularly the very large (2.5 m, 8 ft.) boulders lying in the Vouraikos valley, as reported by (Alevizos & Stamatopoulos, 2018).

The writers here declare that cyclic flooding events, triggered by climate change, characterize the geomorphological history of the valley, and these events were crucial to its formation and "evolution." Some of these "events" were recent, since archeological remains were found during excavations and geological studies, "recent" here meaning anytime within the last 12,000 years, approximately.

They note the presence of alluvial terraces alternating from the north bank to the south bank, which would conventionally mean that a river filled in the valley and then another river washed away the infill by taking a new sinuous course from one side of the valley to the other, right against the valley walls, leaving the terraces as a remnant on the inside of its bends.

The writers then go on to discuss the difficulty in explaining the formation of the strath terraces:

> "It's still not clear the conditions in which the planation of a strath terrace takes place. There is difficulty in deducing the rules that govern the mechanisms of the strath terrace formation along with the duration of such processes and their exact age based solely on field investigation and observations. That is because terrace sequences are difficult to date as they constitute fragmented and incomplete records that were created in climatic, morphological and fluvial conditions that were different from the current one" (Alevizos & Stamatopoulos, 2016).

Apart from the fact that the preceding statement goes entirely against the Principles of Uniformitarian geology, which requires that only processes seen to be in action are to be invoked, the writers go on to invoke just about everything except anything seen to be in action in Greece today:

"Strath and fill terraces have been linked to peak glacial advances . . . deglacial transitions . . . processes not related to glacial processes, like the monsoon intensification or the migration of so-called sediment waves." (I've only heard of the latter in the deep sea, and they're as big mystery there as everything else; that, or they're talking about giant ripples on land, which don't move).

In other words, the writers do not know, or even know where to start. In this case, therefore, the present is of no use in elucidating the past, and uniformitarianism again disappoints.

Following this, they refer to the climatic changes of 11,000 years ago and the shift from cold to warm climate, and then begin to mention floods, flood events, and intense concentrated rainfall, which would, of course, be necessary to provide the water for said floods. Despite having just told us the various difficulties involved in explaining strath terraces, the writers go on to tell us *precisely* how strath terraces are formed:

> "Strath terraces are created when sediment fluxes are high enough to cover the channel bed. Additionally, discharge has to be low enough in order for sediment not to be able to move and expose the bed to be eroded. Finally, lateral erosion has to be greater than vertical incision. High sediment flow that was present during the deglaciation period could easily create such conditions that favor strath formation. In the case of an easily, through abrasion, erodible bedrock, plucking and cavitation, wide straths can be rapidly generated" (Alevizos & Stamatopoulos, 2016).

So, we've gone from not a clue to precisely how they can be rapidly generated, if only conditions are as perfect as possible, and if the rock can be easily eroded, and if the geologist can invoke any agent at all possible or imaginary, like "plucking" for instance. The writers here are saying that the river must flow slowly enough so as not to remove the sediment covering its bed, but at the same time it must be able to erode sideways by abrasion of the rock of its banks by the sediment it's carrying, which, to me, makes little sense.

Apart from the fact that the rocks here at Mt. Chelmos are by no means easily erodible, lateral erosion by a river is generally impossible outside of narrow mountain valleys during times of flood, when the river is heavily charged with abrasive debris. If, somehow, the volume of water in a river increases *permanently* (a very tall order), then the river will now need more room, or else it will simply remain in flood.

A river will form a bigger channel *only if it can*; it will not erode sideways or downwards, beyond enough to accommodate the extra water, and then *only if* the material forming the river bed and the banks is unconsolidated and can be eroded, which is to say washed away, by the river. This issue in by no means simple, and we will meet it again later. We'll now get back to Alevizos & Stamatopoulos (2016).

Like good uniformitarians, the writers go on to tell us that: "Given enough time there is a possibility that 'pediments' may form." Pediments are extra-large strath-like terraces, usually on the lower flanks of a mountain, and which sound very much like small-scale versions of Davis' peneplains. Though strath terraces and pediments are supposedly formed by erosion, they are usually covered by a thin layer of alluvium.

If they are covered, then what was eroding them? The covering is going to protect them, and the water doing the eroding has to be carrying sand and stones, and quickly, to do the actual eroding, since water by itself (or carrying mud or silt) demonstrably, and logically, cannot erode anything, as it has a coefficient of friction of zero.

The foregoing simple physics renders impossible the writers' next statement to the effect that small strath terraces are formed when the river overflowed its banks and "beveled the bedrock" during the increase of supplied sediment. We saw earlier that when a river overflows its banks, only the finest sediment is lifted out of the channel, which, consisting as it does only of mud, silt, and some fine sand near the channel edge, cannot erode anything. Instead, it actually goes to build up natural levees, as we know, thereby covering and protecting whatever is underneath it.

They go on to discuss climate changes and wet periods, and they identify a change in climate coincident with that found by others for the period around 5,000 years ago. Thus, we have the same two prominent changes identified, one at the latter date, and the earlier one at around 11,500 years ago.

They finish out their paper with a brief mention of widespread flooding events that deposited large amounts of sediments in episodes of concentrated rainfall and related flashflood events, which shaped the present-day river system and all the terraces and alluvial fans and other features that we see today.

In other words, these Greek uniformitarian geologists are claiming that the entire area was shaped recently by massive flooding that followed earlier tectonic uplift, which itself had caused the formation of the hard rock morphology, and the recent normal processes simply acted, and still act, on this (Alevizos & Stamatopoulos, 2016).

I must take this opportunity to remind the reader of what this series is all about, apart from Atlantis. The main question to be answered, as I stated at the outset, was whether the earth's surface was shaped by all the currently acting processes we see every day, or whether these are simply acting on a landscape previously formed by some type of catastrophe. And it appears to be exactly the latter that these Greek geologists are saying here.

As usual, they don't give a justification for drafting in more rainfall, and certainly nothing to suggest "monsoonlike" conditions. As we discussed earlier, and as logic would dictate, rainfall in Greece must either be like rainfall today, or less during the Ice Age due to the reduced evaporation to be expected of a colder Mediterranean. Also, they do not define the number of floods, even though such floods should somewhere leave a record of themselves in the form of a discrete layer, or else in the stacked sequences we discussed earlier.

The vague description of flood events, flashfloods, etc., suggests that these writers may be interpreting what is evidence of but a few large and extensive floods, as evidence of many smaller floods, though these smaller floods must still be of significant size. These smaller floods were clearly far larger than any Greece experiences today because not a single writer has compared the evidence of recent and historical floods to the evidence of these large floods from prehistoric times. This, again, makes one wonder about the "present being the key to the past."

Before we finish up with the rivers and valleys of Greece, a brief word on terraces is required. The issue with terraces in any valley is their often-inordinate height above the valley floor, which can only mean that at some period in the past, a flow of water was such that it filled the valley to the height of the topmost terrace. This creates a problem, because, on the face of it, these high terraces take the form of simple riverbanks with floodplains beyond—that's what they look like, and, as Leopold, 1994, tells us (on p. 11) terraces are just abandoned floodplains, or remnants of them. This, of course, would require a massive flood sweeping down through this valley, as we discussed in the Voidomatis.

Since uniformitarian geology does not allow massive floods, an alternative explanation is required for terraces, especially the inordinately high ones. And thus, a combination of Lyell's imaginary wandering river, to first cut out and then fill up a valley, followed by Davis' imaginary tectonic uplifts, has been conjured up to supposedly explain multiple terrace arrangements. In reality, any high terrace is simply a floodplain produced by an enormous flood that filled up the valley to the level of that terrace.

The terrace faces are merely the "riverbanks" of the enormous river this flood constituted, and are formed in exactly the same way as ordinary riverbanks and floodplains anywhere, which themselves are produced by the river as it naturally channels itself to maximize flow efficiency, where the main flow follows the valley thalweg and fills in deadwater areas such as recesses and embayments along its course, again as we saw in the Voidomatis Valley. The valley between the terraces is never filled in and simply represents the (massive) river channel of this flood. The next lower terrace represents a lesser flood, and so on for the other lower terraces.

This, therefore, suggests that these lesser floods may represent returning and declining ebbs-and-flows of the first and main flood, or else later, large floods of lesser magnitude than the first. In which case the stacked terrace arrangement represents a series of major floods and no "tectonic uplift" or valley-filling "meandering rivers" are required. We'll be meeting inordinately high terraces again in the next volume, and what I mean by "high" here is that these terraces are very high up in a very high valley, to the point where one couldn't really get much higher up.

Conclusion

We can see from this and the previous chapter that our dedicated uniformitarians, to a one, seemingly cannot explain the sedimentary evidence to be found all over Greece by any means other than large floods and tectonic action. Another way of putting this, of course, would be to say that at some time in the past, Greece suffered from earthquakes and floods, like some other famous, though mythical, land I could mention.

Plus, the floods on Greece seem to have been very much out of the ordinary, and the tectonic action seems to have been very recent, and significant, and fast-acting, given how recent most of the sediments are. Also, we can clearly see in the present day that sedimentary deposits such as we've been examining here are nowhere found to be forming in Greece, meaning it wasn't everyday processes that produced the older deposits from the end of the Ice Age, or 5,000 years ago.

Chapter References

Alevizos, G. & Stamatopoulos, L., 2016, Landscape Geomorphological Evolution and Coastal Change: A Case Study of Coastal Evolution in the Western Patraikos Gulf Area, Western Greece, *Bulletin of the Geol. Soc. Of Greece*, Vol. L, pp 415-423.

Alevizos, G. et al., 2018, Preliminary Data of Fluvial Geomorphological Evolution and its Link with Hazards and Human Impact: The Case of the Peiros River, North Western Peloponnese, Greece, *Earth Science*, Vol. 7, Iss. 1, pp 11-16.

Caputo, R. et al., 1994, The Pliocene-Quaternary tecto-sedimentary evolution of the Larissa Plain (Eastern Thessaly, Greece), *Geodinamica Acta* (Paris) Vol. 7, No. 4.

Davis, W. M., 1899, The Geographical Cycle, *The Geographical Journal*, Vol. 14, No. 5.

Flemal, R. C., 1971, The Attack on the Davisian System of Geomorphology: A Synopsis, *Journal of Geological Education*, Vol. 19, Issue 1.

Gaki-Papanastassiou, K. et al., 2010b, Recent Geomorphic Changes and Anthropogenic Activities in the Deltaic Plain of Pinios River in Central Greece, *Bull. Geol. Soc. Greece*, Proc. of the 12th International Congress, Patras, May 2010.

Leopold, L. B., 1994, *A View of the River,* Harvard University Press, Cambridge, Massachusetts.

Litchfield, N. & Berryman, K., 2006, Relations between post-glacial fluvial incision rates and uplift rates in the North Island, New Zealand, *Journal of Geophysical Research*, Vol. 111, Issue F2.

Macklin, M. G. & Woodward, J., 2009, River Systems and Environmental Change, in: *The Physical Geography of the Mediterranean*, Ed: J. Woodward. Oxford University Press.

Migoros, G. et al., 2011, Pinios (Peneus) River (Central Greece): Hydrological – Geomorphological Elements and Changes during the Quaternary, *Cent. Eur. J. Geosci.* Vol. 3, Issue 2, pp 215–228.

Pope, R. & Hughes, P. D., 2014, Linking glacial melting to Late Quaternary sedimentation in climatically sensitive mountainous catchments of the Mt. Chelmos complex, Kalavryta, southern Greece, *Geophysical Research Abstracts*, Vol. 16, EGU2014-10905 EGU General Assembly 2014.

Rozos, D. et al., 2006, Large-scale engineering geological map of the Patras city wider area, Greece, IAEG Paper number 241, © Geological Society of London 2006.

Smith, G. W. et al., 2006, Pleistocene glacial history of Mount Olympus, Greece: Neotectonic uplift, equilibrium line elevations, and implications for climate change. In: *Postcollisional Tectonics and Magmatism in the Mediterranean Region and Asia,* Eds: Dilek Y. & Pavlides P., Geological Society of America Special Papers, Vol. 409, Jan. 2006.

Stamatopoulos, L. & Alevizos, G., 2018, Morphological evolution of the urban landscape of the city of Patras and possible natural hazards. In: Abstract book of the SGI-SIMP Congress, geosciences for the environment, natural hazard and cultural heritage, Catania, Italy, 12-14 September, 2018, 753. Roma: *Società Geologica Italiana.*

Tsanakas, K. et al., 2019, Geomorphology of the Pieria Mountains, Northern Greece, *Journal of Maps*, Vol. 15, No. 2, pp 499-508.

Van Andel, T. H., Zangger, E. & Demitrack, A., 1991, Land Use and Soil Erosion in Prehistoric and Historical Greece, *Journal of Field Archaeology*, Vol. 17, No. 4, pp 379–396 (pp 384, 389–390).

Zangger, E., 1992, Neolithic to Present Soil Erosion in Greece, in: *Past and Present Soil Erosion, Archaeological and Geographical Perspectives*, Eds.: Bell, M. & Boardman, J., Oxbow Books, Oxford, The David Brown Book Company, Bloomington, Indiana, U.S.A.

CHAPTER SEVEN:

THE COASTS AND ISLANDS OF GREECE

HAVING FINISHED WITH THE ICE AGE AND ICE action, and with rain and rivers and their valleys, we now move on to the last significant erosional agent of the uniformitarians, the seacoasts, where the mighty power of the ocean is brought to bear on the land. The ocean beating on the land is considered a very effective agent at wearing away a landmass, by the waves pounding against cliffs, and so on.

The waves are also held to have the power to break up the solid rocks into boulders and cobbles and then, by agitating them and banging them against each other, grinding them into ever finer particles to produce shingle and sand, as are commonly seen on the shore. Ocean waves are also credited with the power to sweep sediment away from the land and out into the deep sea. Based on the previous chapters, the reader might, at this point, suspect that I am about to suggest that the foregoing may perhaps not be the case.

The reader would be quite correct; there are as many fallacies and misapprehensions regarding the power and effectiveness of the ocean and its waves as there are regarding the power of ice or rain or rivers, or whatever else. The reality is that the ocean, even at its most violent, has little or no effect on rock, as in hard consolidated rock like granite, sandstone, limestone, etc., due to the fact that those rocks have a compressive strength far greater than the strength of any wave.

If one takes a walk by a seashore with rocks and boulders, one will see that many are covered with seaweed, algae, etc., and hence, like slime-covered river rocks, indicate that the ocean waves have no erosive effect on those rocks. Even on the exposed western Irish coasts, the rocks and cliffs are profusely covered with seaweed and slime, and hence do not suffer any ill effects. Further, the evidence also demonstrates that the ocean has no "long-term" effects on the cliffs and solid rocks either. Even the biggest waves can do little more than shuffle and scatter loose boulders and shingle, etc., about the shore.

The Aegean, on the other hand, is a very shallow sea in comparison to the Atlantic and has a fetch of only a few hundred kilometers versus the 5,000 of the ocean that batters Ireland. While there are certainly storms, the size and power of waves are very limited.

Waves do have an effect on the coast but only on such materials as loose or unconsolidated sand and gravel cliffs and beach deposits, as well as friable or previously shattered rock like some shales and fault breccias and so forth. Waves strongly impacting these materials will have an effect in proportion to the rock's integrity, or lack thereof.

Earlier, when we were discussing the Aegean area, we referred to a paper by G. Anastasakis et al. (2006) on the Myrtoon and other basins. In their introduction, and to give some context, the writers give a general overview of the rocks and deposits of the eastern coast of Greece, fringing the Aegean. The rocks and sediments along the coast are considered by these writers, referring to earlier workers, as anywhere up to 10 my old, being supposedly Late Miocene to Pleistocene in age, with many of Pliocene age, i.e., 2-5 my old.

On the islands of Milos and Kimolos are found Miocene (12 my old) rocks comprised of bioclastic (shells) and partly reef (coral) limestones in some "association" with conglomerates overlain by clastic sediments. Conglomerates and clastic sediments suggest water action that the Aegean is not now, or ever was, capable of, and hence, these deposits cannot be due to any normal marine/coastal processes. These sediments, in turn, are overlain by marine silty clays of supposedly Lower Pliocene age (4–5 my).

We'll assess the stated age shortly but, in the meantime, we'll wonder how those "conglomerates" and "clastic sediments" got to those islands, conglomerates being coarse cobbly deposits, as we're aware, and the (nonspecific) clastic sediments are the same as those we discussed earlier in connection with the basins of the Aegean. Conglomerate, of course, is a (well or poorly, to not at all) cemented, or "lithified" gravel. Such sediments do require, as I have belabored in these pages, lots of water moving quickly to transport them.

Fig. 7-1: Remnant block of extremely coarse, bouldery conglomerate, or mega-breccia to be more accurate, exposed on the beach. Some boulders are bigger than the beachgoers, who are useful for scale. Image: CC-BY-SA-4.0 Int. Photo author: Vihou World.

The authors do not tell us the nature of the conglomerates, but fig. 7-1 shows an inappropriate example from Firiplaka beach. To call this a mere conglomerate is to seriously understate the case, and it was obviously no ordinary "megaflood" that moved this ultra-coarse material. Also, there is no sign of any layering, meaning it was all moved as a mass, i.e., in one action, or had a different origin. The sharpness of the clasts do indicate little to no transport.

Fig. 7-2: The cliffs on either side appear to be composed of different rock types, so the conglomerate is probably fault rock (breccia) minimally transported. Image: CC-BY-SA-4.0 Int. Photo: Vihou World.

Though this mass is isolated in fig. 7-1, fig. 7-2 shows that it may be the disconnected end of a minor headland, or it may, in fact, be a boulder. Judging by the lack of debris around the block, debris which seems ready to fall off due to the obvious poorly cemented nature of the rock, and the fact that the Aegean has very little wave power to either erode or transport, I would hazard a guess that if it was a headland, it was deliberately cut through to interconnect the two beaches that exist on either side of that former headland. On the other hand, to me, it looks more like it just broke off from a mass of the same material in the cliff and fell out onto the beach.

In the Saronikos Gulf in the north, we find the Pliocene marls and sandstones overlain by: ". . . several hundred meters of conglomerate, sandstone, marl and lignite . . ." the latter being brown coal, which is partly coalified wood or peat. This sequence is unconformably (meaning erosion intervened at some point) overlain by 150 m (500 ft.) of ". . . fluviatile conglomerates and sandstones . . ."

The mention of lignites here in the Saronikos Gulf is interesting, considering it implies that large amounts of organic materials are lying very close to the Aegean Sea. In which case, my earlier suggestion of a terrestrial origin for the sapropels in the Aegean would seem to have some support, which suggestion we will revisit when we discuss the Greek lignites later.

South of Athens, on the west coast of the Attiki peninsula, are more fluviatile conglomerates, sands and marls. A marly limestone containing detached and reworked (eroded/transported) blocks of reef limestone, with more limestone on top. This sequence is apparently again unconformably overlain by: ". . . transgressive (invading sea) conglomerates and beach sands." The plain in the eastern part of this peninsula contains: ". . . up to 700 m (2,300 ft.) of reddish terrestrial, mostly coarse clastic sediments . . ." (G. Anastasakis et al., 2006).

Along the east coast is a mass of up to 80 m (260 ft.) of fluvial conglomerates, which are overlain, or pass laterally into, a marine Pliocene sequence. Similar sequences are seen in places all along the coast, overlying the basement rocks. The Pliocene strata have been dated to about 3 to 4 my. Again, they consist of "basal marine conglomerate and sandstone," the marine conglomerate assumed to be of shoreline origin (Anastasakis et al., 2006), which is the usual Lyellian diagnosis of these types of deposits.

However, it should be pointed out that the beach and shoreline gravels and sands along the present-day northeastern coast of the North American continent, and in many other northern coasts, were deposited during and after the Ice Age, supposedly by ice and meltwater floods, and therefore do not have their origin in the usually assumed "breaking up of cliffs by ocean waves." Simply because one sees gravel and sand next to a cliff that waves beat against, does not *necessarily* imply that those waves produced that gravel and sand. That *may* (rarely) be the case, but it does not do to make assumptions, and that is an unfortunate habit the science of geology is notorious for.

Fig. 7-3: Beach deposits on Santorini clearly showing the distinct differences in rock types between the stones and cobbles on the beach (all grey, black and red) and the pale-colored rocks of which the cliffs are composed. Clearly, therefore, based on the color alone, the beach deposits did not derive from the breakdown of the cliffs, and it is highly questionable how they got where they now are, given that the Aegean, as it is at present, could not possibly have done more than shuffle them about. Their rounded to subrounded shape, however, tells us they are well-travelled. Photo by Andres M on Unsplash, free license.

In the Aegean Sea, for example, waves have little power, due to the limited depth and fetch, or amount of open sea, that wind can work on to create waves of any size or strength. And, all we *do know* about the geological history of Greece and the Aegean tells us that, at present, the Aegean Sea is about the biggest it's ever been.

Hence, one is required to question the origin of the conglomerate and not assume it's a basal "marine conglomerate" just because it's near limestone, near the edge of the sea. Since it's on bedrock, we can simply call it the "basal conglomerate" that it is, and nothing more—especially considering basal conglomerates are found everywhere, often thousands of miles from the ocean.

On the island of Aegina are more terrestrial conglomerates, overlain by marls and marly limestones. At Agios Thomas, an island to the northwest of Aegina, is a thick succession of sediments capped by an andesite breccia (Anastasakis et al., 2006), a breccia being a conglomerate composed of sharp angular fragments of rock typically with some fine material, such as the mass on Firiplaka beach shown in fig. 7-1.

Andesite is interesting in that it is an extremely hard igneous rock derived from the volcanic eruption of melted continental-type rocks, and is very widespread in the Andes, hence the name. Brecciation of andesite implies a very powerful, and necessarily sudden, disturbance. A slow-acting process, such as the theoretical plate-tectonics, wouldn't do this—one shatters with a shock.

As stated at the beginning of this section, the rocks in question were all thought to be from 10-2 my old, mostly Pliocene in age. The The Pliocene epoch spans from about 2.5–5.5 my ago, and is the last division of the Tertiary period, as defined all those years ago by Lyell himself. Further, it used to be the last epoch immediately prior to the present age, what we call the Holocene, the time since the Ice Age.

However, according to that same English academic who imparted all that valuable uniformitarian "knowledge" to me all those years ago, Lyell decided to interject the almost 2.5 million year newly invented Pleistocene epoch in between. He did this because the fossil molluscs of his Pliocene were too similar to those of the present day, and thus his theories of vast ages required him to push his Pliocene epoch further back in time (conveniently for himself) and assign the modern-looking molluscs to his newly invented "Pleistocene" epoch (Whitcomb & Miller, 1984).

Furthermore, according to the same man, Lyell also needed lots of time to allow all the Pleistocene megafauna to go extinct, species by species, in the slow gradual way his uniformitarianism called for. This, of course, was mere presumption on Lyell's part as the concept of mass extinctions was not in his conservative philosophy, it being an utterly forbidden catastrophist notion, à la Baron Cuvier and many others.

Given that we now know the megafauna (mammoths, mastodons, and so on) were killed off in a mass extinction event at the end of the Ice Age, we don't need any invented Pleistocene epoch to allow for any such nonexistent gradual dying-off as per Lyell's personal concepts of "whole species" lifecycles, referenced earlier in connection with W. M. Davis. In which case, we now have grounds for a certain suspicion regarding his timescale. The Pliocene may, in fact, be much more recent than Lyellist geology claims and this brings all other ages into question also.

As a matter of interest, William H. Dall, of the U.S. Geological Survey, reported Pliocene molluscs in unconsolidated beach gravels near Nome, Alaska. He dated them as Pliocene based on Lyell's percentage method, which says that any deposit with more than 10%

of extinct species is older than the Pleistocene. Dall's contained 37% extinct species, hence Pliocene (Dall, 1920).

Interestingly, he tells us that many of those species are alive and well and living on Cape Cod, Massachusetts, at this very day, and he does wonder, as might anyone, how the exact same species of mollusc could exist on both sides of the American continent, 5,000 miles apart and no connection between them or means of travel. He further reports that F. C. Schrader, also of the USGS, found Pliocene fossils in unconsolidated silts near Colville (cited in Dall, 1920) which is also in Alaska.

Much more recently, T. A. Ager et al. (1994) reported plant fossils from the Yukon River near the village of Circle in east-central Alaska. These so-called fossils were found in sandy silt lenses within gravels and included seeds, pinecones and needles, and pollen from trees and shrubs, many of which still grow there today, and many that don't. These plant remains are not actually fossilized, they are simply preserved pine needles, cones, etc. The authors date these undecayed remains as Pliocene, based on similarities with other remains found 250 km to the south, in gravels considered to be of Pliocene age.

Seeds, pinecones, and pine needles lying in loose gravels, and all undecayed and unfossilized after three or four million years is more than a bit questionable.

One has to wonder, therefore, just how long ago the Pliocene epoch ended, given that 3 million years is a long time for pinecones and pollen to survive. The shells found by Dall and others have been lying in loose gravels, sands, and silts for just as long, while the gravel has not consolidated and the shells have not disintegrated or been weathered away, even with the cold, and hence CO_2-heavy, rainwater infiltrating through them all this time.

Not only that, but an ice sheet is supposed to have bulldozed its way over the entire country not more than 12- or 20,000 years ago. How was it that a supposedly all-destroying ice sheet glided so gently over all these gravels? There is no way that I can possibly accept an age of 3 or 4 my, or anything like it, for this undecayed organic matter. Therefore, we may go back to the Aegean and ask whether all those conglomerates, and other rocks and deposits, are not perhaps much younger than geology says they are.

The evidence from the Aegean certainly suggests this to be the case, with the sea's almost complete lack of any ability to erode or transport today. But still, those poorly consolidated conglomerates are present, just as Alaska has its molluscs and undecayed plants. It is clear, from their nature and coarseness, that those conglomerates, at up to 80 m (260 ft) thick, and no matter when they were deposited, are the product of large-scale water movement and debris transport, what we normally refer to as massive, or catastrophic, flooding. Hence, while it's not possible to say for certain, these conglomerates are very likely the products of the same floods that swept material from the northwest and filled in all those basins in the Aegean. That, or they're the result of some earlier flood.

Working in the Peloponnese and Crete, R. Paepe et al. (2004) describe ". . . thick series of gravel beds, each overlain by sands and loams . . . thin series of sand, loams, gravel . . . topped by thick gravel bed . . ." and many such sequences. These very much sound like flood deposits, just like the typical upward-fining sequences we discussed earlier. They identify multiple cycles which they link to Milankovitch (Ice Age) orbital cycles of 100K years each, and report that at the end of each cycle, a "huge" gravel bed is deposited, for which they do not offer much in the way of an explanation, other than the very vague suggestion that they are related to:

"phases of drought and cooling off," which would suggest some kind of polar desert.

They then claim that such huge gravel deposits can only form by steady rock fragmentation under desertic conditions. Apart from the notion of a "polar desert" near the Mediterranean being more than a bit far-fetched, rock fragmentation occurs under different conditions, most usually frost action and thermal effects in hot deserts, but in a polar desert, where the temperature remains perpetually below zero, as in the Antarctic interior, *absolutely nothing* goes on.

This claim of a polar desert is highly questionable in view of frost action, which requires liquid water—we've seen vast amounts of gravel in the river valleys of Greece and no hint of a polar desert. A hot desert doesn't really "cool off," so I'm not quite sure what these writers are driving at here. Huge gravel deposits, obviously, can be formed by floods, or by glaciers as moraines, as we've seen, so their claim is a mystery to me. Thermal effects in hot deserts produce fragments that get washed away by flashfloods in wadis, and so also produce gravel deposits, but there is no hint of a hot desert in Greece or Crete, and while there are dry valleys, no one called them wadis. These "huge" gravel beds, much as we've met throughout Greece are obviously, and simply, flood deposits like all the rest.

Finishing up on Crete in the south we'll examine an archaeological perspective from Runnels et al. (2014), investigating Pleistocene maritime activity in the Aegean. They begin by exploring an alluvial fan on the north coast that is 4 km wide and extends 600 m inland. They tell us that it is Quaternary, which is to say Pleistocene rather than Holocene, and it was created by:

> ". . . large magnitude sedimentation events within local watersheds through extensive erosion of upland mountain hinterlands to the south that produced vast volumes

of boulders, cobbles and coarse-grained sediments, paving valley floors and accumulating in alluvial fans. These probably climate-forced, high-volume flows are interrupted periodically by extensive over-bank deposits of finer-grained clastic sediments, formed perhaps during periods of climate stability. Fan development may have occurred during climate pulses, with intervening periods of stability that allowed vegetation growth and soil development. The fans graded out at times of glacial sea level low-stands to a now-submerged shoreline several kilometers offshore. Today, they terminate at the shoreline where steep scarps are being created" (Runnels et al., 2014).

The authors here, I must say, are relying a bit much on such words as *probably, perhaps,* and *may,* words that certainly do not imply any kind of certainty or its approximation.

They also appear to be doing everything possible to avoid the word *flood,* and follow Zangger et al. with their use of the term "sedimentation event." They go a lot further, however, and have introduced the term "climate pulse," which either means nothing or anything one wants it to mean. I've never heard the term until now.

In reality, there is no such thing as a "climate-pulse," or a means to rapidly change climate, or have it deviate sharply or temporarily from a norm; such a thing is simply a bit of bad *weather*, always very short-lived and minor, and yes, we may have a wet year or two at times, but no one has ever reported what the authors here suggest. (We're ignoring the Ice Age, as it's a totally different concept and long-term in any case, and the "climate change" we're now experiencing is, strictly-speaking, a minor deviation from some poorly defined norm).

What the authors are implying here, however, is a major step outside the norm, with their "vast volumes of boulders" and "alluvial fans," all of which imply major flooding events, just like everywhere else we looked in Greece. These "alluvial fans," as described by the authors, sound suspiciously like the alluvial fans of Nemec & Postma (1993) whom we met earlier, and who considered these fans to be deposits from meltwater floods coming off proposed glaciers on Crete, which no one bought.

Those glaciers were also held responsible for the boulders scattered over Crete and choking up many of the narrow mountain valleys, as Paepe et al. (2004) confirm. Two major fans dominate in the area, and they are (logically) related to the two biggest drainage areas.

The fans, therefore, cover an area of 4 km by about 3 km, and extend well out from the mountains in which they originated. Sea level may well have been lower, but a powerful flood such as could move this material would not be stopped at the water's edge; it would simply blast that water out of its way, as can easily be imagined.

At the same time the use of the phrase *sea level low stands* seems to be the usual effort to stretch the deposition of these fans out in time, all uniformitarian-like, as if they took tens of thousands of years to put down, spanning multiple glacial and interglacial cycles alike, when it seems clear, since they're on top of everything, that they were deposited at the end of the Ice Age, as Nemec & Postma (1993) claimed, or else they're even later. The latter are, at least, geologists rather than archaeologists, even if they are card-carrying uniformitarians.

In describing the fans in more detail, the writers tell us that they are composed of a series of layers, each 2-3 m (6-10 ft.) thick, of cobbles and boulders interbedded with finer-grained deposits, sounding similar to the layered deposits shown in fig. 5-5, also from Crete. This stratigraphy, for some reason, the writers interpret as boulders and cobbles representing "outwash," while the finer-grained material was deposited during periods of stability.

This does not make a lot of sense, as the outwash implies a flood to move the boulders and cobbles, and floods move more than just boulders and cobbles. The sequences sound like typical flood sequences, as usual, coarse at the base and grading upward into finer material, as we've discussed multiple times herein.

The situation here on Crete seems to be no different, the floods here appearing to have come from the south. And, if they came from the south, they must have come *over* Crete because there isn't now and never was any source of large volumes of water on Crete itself, and large volumes of water are needed to move large boulders, while ultra-violence is needed to create them in the first place.

The writers report finding a number of artifacts in the fans, so humans were obviously around, and thus, the fans couldn't be too old. From the description of a section of one fan, it seems that little in the way of thin layering is apparent; rather, it seems that the section shows two "macro-layers." A paleochannel filled with cemented conglomerate is found at the base, (another "basal-conglomerate"), and this is covered by a thick (3 m, 10 ft.) layer of what they consider paleosol (fossil soil) which contains stringers of gravel and small stream channels filled with "debris-flow" material.

This sounds to me like the braided channels one sees on the surface of alluvial fans, due to the declining flow of a flood during its final drainage. Because of the time needed for soil formation, and the mere 12,000 years since the Ice-Age, the 3 m/10 ft. of paleosol cannot possibly be soil that developed in place since then; it is without question the usual fine silt and mud deposited as any flood declines, and which plants then colonize when the flooding stops.

While the foregoing is extracted from relatively recent papers on Crete, we also have a report from British Royal Navy Captain Thomas Spratt, from 1865. Spratt was engaged in hydrographic survey work in the general Aegean and Black Sea regions, so he was something of an early oceanographer. As was standard practice in those days, he also conducted geological surveys of the neighboring territories, in southern Russia, Bessarabia, and the regions around the Aegean and Black Sea.

From the geology, he concluded that the whole region, up to recently, was covered with lakes, since only freshwater deposits (marls) and fossils are to be found. This, as we saw earlier, was confirmed by Suess (1904). Spratt concludes that freshwater lakes extended all the way through the Black Sea, the Caspian, the Aral Sea and on to the Urals, all of which he thinks were probably connected during this "Lacustrine Period" (Spratt, 1865 p. 365). He is here talking about the Paratethys we met in connection with Hesiod's *Theogony* in volume 1.

He notes that in the northern part of the Aegean Archipelago there are two series of freshwater deposits that lie unconformably on one another. This is what we have seen more or less everywhere in Greece, i.e., series of deposits (units), breaks in deposition, and with erosion surfaces in between each unit, which can only be explained by positing a series of large floods. Spratt also mentions brackish-water deposits but no recent marine deposits anywhere.

Like much of the rest of Greece, the more recent deposits on Crete are found in-filling basins. Spratt reports that such deposits consist of gravels, sands, and marls, which he dates as "Newer Pliocene" (Lyell's term) which is to say Pleistocene. And, as with all Pleistocene deposits, it is connected to the Ice Age somehow or other, usually being deposited at the end, like most other Pleistocene formations we've seen.

When it comes to the recent deposits, Spratt reports that the most recent, and overlying everything else, though patchy in distribution, is a bed of "red earthy gravels of a late date." These gravels are scattered throughout the interior of the island and occur at various elevations. He notes here that they are not of purely aqueous origin, nor were they deposited in a tranquil state (Spratt, 1865, p. 359). The latter is not surprising, since, as we all know by now, tranquil water cannot transport or deposit gravelly material, or much of anything else for that matter.

Spratt, however, considers that these red gravels deserve special attention because of where they occur, it being somewhat difficult to account for their distribution. Here we have echoes of the ophiolites of the Voidomatis Basin. The red gravels, Spratt tells us, occur in most wide valleys and at various elevations but are localized rather than generally dispersed. They are always found uppermost, implying they are the latest materials to be laid down.

Their origin is uncertain, but Spratt considers that they indicate a drift-influence (moved by something) from the irregular arrangement of their beds, which can be nearly 100 ft. thick (30 m). They contain large boulders, as Nemec & Postma, and Paepe et al., reported, and Spratt thinks that something other than water action must be required to account for their thickness and the locations and positions in which they're found.

The issue is that the gravels contain certain schistose boulders 3–4 ft. (1 m) in diam. and thus must have come from the schistose ridges to the south, at least four or five miles (8 km) away from where he found them. They must have travelled over and across the intervening ridges and valleys, since the earthy gravels also occur on the crests of the schistose ridges and on other ridges of marls and other rocks.

Spratt concludes, from the localized occurrence of the gravels, that their dispersal was also a more localized process than a more general one such as a "wave of translation," and thus he invokes ice in conjunction with water. In support, he claims that some of the gravel deposits are in the nature of moraines, or "like a moraine" (Spratt, 1865). We know already that that doesn't mean much, as the various types of deposit are easily confused.

However, Crete is much too far south, Ice Age or no Ice Age, to have ever seen so much as an icicle, let alone a glacier, so I can't give Spratt's ice suggestion any credit. As noted, this is much the same issue that we had with the ophiolites of the Voidomatis River valley. Here on Crete, the boulders demonstrably came from the south, over ridges and across valleys, and only a massive flood could do this kind of transporting, the same flood, from the south, that no doubt formed Runnels et al.'s massive alluvial fans we looked at above.

Spratt finally notes that he found the same red earthy gravels and boulders in the Melaxo Basin above the town of Khania, at an elevation of 1,800 ft. (550 m) three ridges north of the schistose rocks. Since ice cannot travel uphill, it couldn't have brought boulders from the south down into a valley and then back up over a ridge multiple times to get red earthy gravels, with boulders, high up into the Melaxo Basin.

Therefore, ice had nothing to do with these red earthy gravels, or their boulders, or the transportation and deposition of them, anywhere on Crete. Spratt, however, ascribes them to some terrestrial or atmospheric condition, rather than the aforesaid "wave of translation." He does admit, though, that considering Crete's southerly latitude, the influence of ice was "probably not only feeble, but intermittent, according to the season's intensity" (Spratt, 1865).

We will meet Spratt again in the Mediterranean and North Africa, and we'll meet the red soils again in later chapters herein.

Conclusion

When we examined a sampling of the coastline of Greece, its islands, and Crete in particular, we saw that the most usual types of sediment were conglomerates, coarse gravels, sands, and even boulders, all but the sand and pebbles being beyond the feeble powers of the present-day Aegean to so much as shuffle about. We also saw that the beaches are littered with stones and cobbles unrelated to the local rocks, and we noticed that almost every writer on the subject had to come up with some unseen, unknown, and highly questionable process or mechanism to account for them.

As I noted in various places herein, my proposed solution of a major, overwhelming flood would quite easily account for all of them, just like everywhere else. And thus far, no one else has come up with anything in any way demonstrably competent.

It would seem therefore, that uniformitarianist geology cannot explain the geological evidence we have thus far seen in most parts of Greece and the Aegean that we have briefly examined. By "briefly" I mean that many decades of work involving PhD students, post-graduate researchers, and academic and industry professionals have gone into the study of some of these mountains and river valleys, and many other study areas of very limited extent, while I have only extracted a few pages worth from what could comprise a library on many of the topics.

Obviously, from the establishment point of view, and from what results, or lack thereof, we've seen herein, all these PhD studies and all other research have been done within the paradigm of uniformitarianism, which appears generally incompetent. Thus, little but theory and supposition has resulted.

At the same time, while trying to maintain a brave uniformitarian face, most researchers, even the British ones, have had to admit that the evidence indicated repeated flooding events, but they have at all times downplayed any significance, and done their very best to shoehorn the evidence into their precious paradigm.

Mind you, and as we'll see further, the native Greek geologists were not quite so reticent about flooding events, even if they still purported to adhere to the dictates of imported British so-called science. We have examined the recent geology of Greece closely enough for our purposes, and we have found that, without question, British uniformitarianism is found seriously wanting when it comes to explaining that same Greek geology.

Quite distinct from any kind of slow, gradual, everyday processes indicating their responsibility for the evidence, that evidence everywhere indicates that major and rapid water movements occurred on a large scale, and did so repeatedly, all over the country. Even the orthodox geologists we consulted were pretty much all constrained to admit of least two, and usually three (if not actually four) major flooding events as having occurred on Greece.

Even Spratt suggested it for Crete merely by mentioning it in the negative. Considering when he was writing (1865) and that he had a career to consider, this was very likely Spratt being careful to follow the lately established convention of uniformitarianism, rather than the catastrophist view commonly held up to this time, and we'll see that Spratt himself had a very different attitude 25 years earlier.

We have also seen much clear evidence for subsidence, especially in the Aegean Sea region, along with plenty of apparently recent evidence for tectonic movements and faulting, etc., in the basins and the river valleys on the mainland. I must note again here, however, that it is also somewhat coincidental that evidence for three or four Ice Ages or ice advances were also identified; I'm inclined to suspect, as I suggested earlier, that perhaps deluges of water, and not ones of ice, were the more likely culprits.

Thus, we seem to have some confirmation from establishment geology that three or four large floods struck Greece not too long ago, which, along with the evidence for subsidence in the Aegean, seems to support the tale of the old Egyptian priest. At the same time, it is impossible at present to tell if the evidence implies three or four separate floods or simply one that ebbed and flowed a number of times. At least however, we have seen evidence of at least two time-separate major flooding events on Greece, one at the end of the Ice Age and a lesser one 5,000 years ago. We will now move on to a whole other branch of geology, the organic, and see what it has to tell us of the relatively recent geological history of Greece.

Chapter References

Ager, T. A. et al., 1994, Pliocene terrace gravels of the ancestral Yukon River near Circle, Alaska: Palynology, paleobotany, paleoenvironmental reconstruction and regional correlation, *Quaternary International*, Vol. 22-23, Issue C, pp 185-206.

Anastasakis, G. et al., 2006, Upper Cenozoic stratigraphy and paleogeographic evolution of Myrtoon and adjacent basins, Aegean Sea, Greece, *Marine and Petroleum Geology* 23, (2006) pp 353–369, pp 363, 365.

Dall, W. H., 1920, Pliocene and Pleistocene Fossils from the Arctic Coast of Alaska and the Auriferous Beaches of Nome Sound, Alaska, USGS Professional Paper 125-C, Washington Government Printing Office, 1920.

Nemec, W. & Postma, G., 1993, Quaternary alluvial fans in south-western Crete: sedimentation process and geomorphic evolution, *Special Publication of the International Association of Sedimentology*, Vol. 17, pp 256-276, cited in Hughes et al., 2006.

Paepe, R. et al., 2004, Quaternary Soil—Geological Stratigraphy in Greece, *Bull. Geol. Soc. Greece*, Proceedings of the 10th Intl. Cong., Thessaloniki, April 2004.

Runnels, C. et al., 2014, Paleolithic research at Mochlos, Crete: new evidence for Pleistocene maritime activity in the Aegean, *A Review of World Archaeology*, Vol. 88, Issue 342.

Spratt, Capt. T. A. B., 1965, *Travels And Researches In Crete,* Vol. II, Appendix IV: On the Geology of Crete, London, John Van Voorst, Paternoster Row.

Suess, Eduard, 1904, *The Face of the Earth*, (*Antlitz der Erde*).

Whitcomb, N. J. & Miller, W., 1984, Lyell's Proposal of the Term "Pleistocene," *Tulane Studies in Geology and Paleontology*, Vol. 18, No. 2, pp 77-81.

CHAPTER EIGHT:

ORGANIC GEOLOGY
AND PALAEONTOLOGY

THUS FAR WE HAVE FOCUSED ON THE GENERAL, I.E., inorganic, geology of Greece and the Aegean, with little mention of fossils or plant remains. A major feature of the geological structure of the region is the multitude of grabens, or basins, mentioned through-out this study. It is interesting to note that of all the basins of Greece, whether structural, which most of them are, or other types such as simple mountain valleys, over 60 contain coal deposits, as well as other kinds of deposits, even more unexpected.

These "coal" deposits actually comprise anything from peat and lignite (brown coal) to subbituminous coal (low-grade black coal) with the majority being lignite. The lignite deposits are multi-lay-ered and are considered by academia to have formed intermittently over the usual long periods of time, in this case from the Miocene to the end of the Pleistocene, or over the last 20 my or so, and to have formed in mires or bogs that developed in these basins.

I find it more than curious that the vast majority of these basins, over the whole extent of Greece, all display a very similar profile. It's as if, on and off over the long course of 20 my, all these basins, in very different environments, different geological settings, in different regions and at very different elevations, just happen to share an almost identical history. To my mind, it is a bit hard to accept that over 60 very similar phenomena, spread widely over such a large area (and with many more in surrounding countries), could each have been created by the occurrence of exactly the same conditions at different or overlapping times.

Not only that, but exactly the same changes, or very similar ones, involving supposed subsidence and sediment deposition, also occurred simultaneously in all these basins. It seems logical to me to conclude that when we find the same thing occurring at the same time in many different places, it is due to the same overarching cause, as Occam's sensible Razor would suggest.

Many of these basins are at rather high elevations in the mountains, up to 800 m (2,600 ft.). This is somewhat unusual, as mires, swamps, and bogs tend to form in low-lying poorly drained areas, not in mountainous areas, which are typically well-drained due to the slope, or else not drained at all, in the case of an enclosed valley. In the latter case, an enclosed valley will naturally fill with water to form a mountain lake. A mountain lake does not turn into a bog unless it gets filled up with something, which is a great deal harder to accomplish than it sounds, but lakes filling up and developing into bogs, as we'll see, is the explanation for all the lignite deposits, no matter where they are.

Further, these basins typically contain multiple layers of lignites, separated by layers of different sediments, including marls, sands, clays, and even conglomerates, which means, on the face of it anyway, drastic changes of conditions occurring at various times, and the sands and conglomerates especially, and as we've seen, imply high-energy conditions, which are harder to envisage in an enclosed mountain basin than multiple layers of lignites

The answer offered by academia is that any basin, wherever it was, slowly subsided over long ages, and as it did so it was synchronously filled in, the subsiding and filling balancing each other nicely to result in a typical flat- (or slightly uneven) floored basin just like the ones we discussed earlier, all over Greece and in the Aegean—the latter, as we know, having been dry land with freshwater lakes for much its sub-recent history.

One thing that does make me wonder, though, is that if these basins have been subsiding and filling over such a long time, why has no lignite been found by drilling in any of the basins of the Aegean? We discussed, in some detail, the structure and infill of these basins earlier, and no lignite is mentioned by anyone and a search turned up nothing.

If all these basins on the Greek mainland could produce multiple lignite deposits, substantial enough to be mined for power generation, then why couldn't the same types of lake basins in the Aegean region have produced them? They were, according to our sources—and Spratt is hard to gainsay—no different for similarly long ages, the Aegean landmass having only recently subsided.

We do allow (somewhat reluctantly) that the Aegean may have previously been submerged for certain periods in the distant past. The evidence indicates, as Suess (1904) and others have stated, that Aegeis was mostly above sea level since before the Pleistocene, i.e., for the past 3 or 4 my, and was only submerged about 12,000 years ago. A few million years is surely plenty of time to fill in a lake, form a bog, build up peat, and bury it under more sediments to compress it into lignite, just like on the now-mainland of Greece and everywhere else. Yet no lignite is reported from any drilling in the Aegean.

The lignite formation process begins with peat development. I. M Kurbatov (cited in J. P. Andriesse, 1988) summarizes 35 years of research into the formation of peat as follows:

"The formation of peat is a relatively short biochemical process carried on under the influence of aerobic micro-organisms in the surface layers of the deposits during periods of low sub-soil water. As the peat which is formed in the peat-producing layer becomes subjected to anaerobic conditions in the deeper layers of the deposit, it is preserved and shows comparatively little change with time." Andriesse (1988) later continues: ". . . the presence of either aerobic or anaerobic conditions decides whether any biomass will accumulate and in what form.

Distinction is made by Kurbatov (1968) between forest peat, which is more aerated and therefore more decomposed, and peats [that] formed under swampy conditions with strongly anaerobic conditions. In forest peat, lignin and carbohydrates appear to be completely decomposed so it generally has a low content of such organic compounds, whereas under swampy conditions, peats are characterized by high contents of cutin and the presence of much unaltered lignin and cellulose. Actually, Kurbatov's forest peat is much the same as thick (forest) litter deposits."

Given that forest litter contains no lignin, it can, therefore, have nothing to do with the formation of *lignite,* and hence all the peat, lignites and coal in the Greek basins must (presumably) come from swamp peat created under anaerobic conditions. Peat often contains all manner of plant debris such as wood fragments, twigs, branches, even sections of tree trunks as well as whole ones, along with pollen, leaves, and other plant matter. These allow identification of the plant types that were incorporated into the peat, and the type of climate and environment which pertained at the time of the development of the swamp and peat.

Plants are excellent, which is to say, the best and unequivocal, indicators of past environments. This is generally because all plants have a rather limited growing range, defined by climate and amount of daylight, which equate with latitude. Ranges vary from narrow to broad, but plants do not grow outside their ranges, so fossil plants will always indicate the climate prevailing while they lived, and where they lived, which, of course, is not necessarily where they are later found, since they float.

All the lignites in the Greek basins contain abundant plant remains, and various workers have identified many of the species and tried to elucidate the climate history and environments of their formation. However, and with reference to the use of the word *presumably* in parentheses above, there is another way to create lignite and coal that does not involve a swamp-peat stage. Despite the process being well-known and observed, it is rarely mentioned.

If, for instance, a mass of plant material, like trees, bushes, etc., were swept away by a flood, such as happens regularly at present, then these would be carried, floating on the water in the form of a mat, or raft, composed of tree trunks, branches, leaves, seeds, etc. If this material is then deposited as a unit in a basin or hollow in a floodplain and buried by subsequent floods carrying fine sand, silt, and clay, etc., then the plant material will, just like peat in any swamp, be compressed in an anaerobic environment and turn into lignite, and thence to coal eventually (as Lyell, 1830, discusses on p. 187 *ff*, with respect to such mats of trees brought down regularly by Mississippi floods).

However, this explanation is generally shunned due to its association with the dreaded concept of floods, which is rather ironic in the case of Greece, seeing that so many geologists resorted to just that concept to explain so much of the geology we've already looked at. At the same time, the difference between a peat-derived lignite and one derived from a flood deposition "event" is easily discriminated.

Peat supposedly grows in place and hence a lignite bed so derived should be underlain by a soil horizon. Further, roots from the peat plants should extend down into that soil. Even though the mass is compressed, those root traces should be identifiable in that fossil soil.

When peat, or some other mass of plant matter, is converted to lignite, the compaction ratio has been determined to be generally of the order of from 2:1 to 4:1, and sometimes more (Widera, 2015). Such a compaction ratio will not obliterate the fossil roots, since coal is estimated to have a compaction ratio, depending on the grade, of the order of from 11:1 to 60:1 (Widera, 2015) and leaf traces are readily observed in at least the lower grades of coal. As far as the Greek lignite deposits go, therefore, the presence or absence of a soil layer beneath the lignite will tell us a lot.

Lignite, as I've said, covers the spectrum between peat and bituminous coal, being the transition stage between one and the other. While there is not a lot of peat or bituminous coal in Greece, there is a large amount of lignite, much of it almost coalified. The lignites of Greece present an opportunity to introduce this branch of geology, especially as the peat bogs of northern Europe, and even more especially, the blanket bogs of Ireland on the Atlantic coast, as well as those of other areas of northern Europe/Russia, bear on the history of the Atlantic, and the question of Atlantis. These blanket bogs are distinctly different from the typical low-lying swamp-bog that develops in a basin, or river valley floodplain, or any poorly drained area.

The word *bog* comes from the Irish word *bogach*, and Ireland's peat bogs are famous for a number of reasons, not the least of which being the many bodies that have been found in them. Many other things have also been found in these bogs, such as ancient artifacts, weapons, books, and so on, but some really interesting things are those that have been found *beneath* the bogs, such as roads, trackways, houses, farms (with furrows still intact), stone walls, stone structures, as well as fossil animals, particularly the giant Irish deer.

In contrast to the Greek lignite deposits, all of which are in basins, many of the Irish peat bogs are actually up in the mountains, and cover the hilltops and slopes, areas where one should not really expect peat to develop, since swamps typically do not form on mountainsides. This is naturally due to the fact that said hilltops and slopes tend to be very well-drained by gravity. The theory of their formation is that minerals such as iron were washed out of the upper soil layers and down into the pre-existing, or underlying soil, where they formed an impenetrable layer known as "iron pan," or just "hardpan," causing the slopes to become waterlogged and the bog to form.

This seems a weak enough effort, because it implies that for some reason the rainwater couldn't just drain away, even with all that slope and gravity to help it. Also, such an explanation is entirely at odds with the readily observable evidence of water flowing out at the base of every bog where it has been cut away for fuel, as shown in the following photograph of a blanket bog, fig. 8-1, from Connemara, Ireland, and I've never heard anyone mention "hardpan" beneath an Irish bog. And, in fact, where I have observed water flowing out of the base of a bog, it did so over a substrate of coarse gravel up on a hillside, however that gravel got there.

Fig. 8-1: Cut face of peat bog with peat bricks arranged to dry. This is a good example of a blanket bog draped over a sloping landscape. Very obvious is the water at the base of the peat-face which, because there is no flow over the bog's absorbent surface, can only have come from the peat itself. Image: ID 11066565 © Garya, Dreamstime.com. Royalty-free Lic.

Quite apart from upland bogs, the fact is that most swamps that exist today show no sign of turning into peat bogs. To my mind, these upland "blanket bogs" should not exist because no circumstance I can think of, or any that have been proposed, can cause the top or side of a mountain to become waterlogged when the water can simply flow off of it due to gravity. In any case we'll leave this Irish mystery for later and turn to the Greek one.

Many of the bigger lignite and bituminous coal deposits of Greece have been exploited for power generation, and thus have been well studied.

This is especially true of those in the north of the country, in the Florina-Ptolemais-Servia Basin complex, which consists of a number of interconnected/adjacent basins containing notably substantial lignite deposits. There are a number of sub-basins within the complex where the lignite beds are particularly thick and thus most economically exploited. According to Mavridou et al. (2003) the Amynteon-Ptolemaida is the most important Neogene basin in Greece because it contains numerous exploitable lignite deposits.

The basin complex is about 100 km (62 m) long and 20 km (12.5 m) wide and is situated at an elevation of 600 m (2,000 ft.) above sea level (a.s.l.). The sedimentary sequences of both sections of the complex are the same and both have been disrupted by tectonic events during the Quaternary (Pleistocene to recent). The lignite deposits of Anargyra (Amynteon) are part of a thick succession and are covered "... in their entirety..." by Quaternary fluvial and limnic (lake) deposits.

Thus, the basin got filled in somehow over an undefined time period, and then during the Pleistocene (Ice Age) something fluvial deposited sediment over the entire basin, just as we've seen with many river basins and coastal plains earlier. The one question that might arise here is how it could be possible for what seems to be an enclosed basin (see fig. 8-10) to have a (fluvial) river running through it capable of depositing fluvial sediments all over said basin.

Underlying the deepest of the lignite beds is a sequence of clastic sediments, comprised of sands and conglomerates, gravels, etc., as well as marly sands, sandy clays and unconsolidated conglomerates, i.e., a sequence of water/mud-transported and deposited coarse material. The lignite is thicker in the northwestern part of the basin and thins out toward the south-east, while it is missing in some places.

Interestingly, the writers tell us that there are all of 27 lignite beds in the basin, which, according to convention, means that a swamp developed, peat developed in the swamp, and then this swamp changed to a lake, was buried by sediment, upon which another swamp developed, etc., 27 times over.

To me, that is really stretching credulity; to expect the precise conditions for swamps to develop and peat to form over and over again for 27 cycles is far-fetched to say the least, especially when we bear in mind that subsidence, tectonic activity, erosion and deposition were all supposedly going on throughout this supposedly 20 my period.

Plus, I don't think it can seriously be argued that the tectonic movements could possibly have been so finely calibrated as to produce such perfect step-wise, up-and-down adjustments to enable such neaty-peaty repetition. But, here gain, a "leap of faith" and a decent-sized wheelbarrow of salt are all that are required to swallow this "theory."

Of note is that the peat (lignite) layers and the interbedded sediment layers are supposed to represent completely different environments, but the same fossils are found in both. And, the types of fossils present raise some questions. In the lignite, the most common fossils are the gastropod called *Planorbis melanopsis,* a freshwater (lake) snail, not a swamp snail. Another is *Paludina,* a pond snail and *Theodoxus macedonicus,* also known as *Neritina.* The latter lives in both fresh and brackish to marine water and needs stones and a rough stony or sandy bottom to live, which is more typical of a river or seashore. It is questionable as to how it got 600 m (2,000 ft.) up into the Greek mountains to find itself in a swamp and incorporated into a bed of lignite.

Obviously, it didn't get there under its own steam, and, like every other type of deposit, it must have been transported by something from its natural environment by the shore, swept up into the hills and deposited in the basin where we now find it. *Paludina*, the pond snail, also has little business in a swamp. However, snails such as *Paludina* fill cavities with air and float to new locations during floods, so its presence here can be explained, whatever about the flood that brought it.

Based on data collected from 10 drill holes, the mean thickness of the lignite is 30.75 m, ranging from 17. 5-48 m (100 ft., ranging from 60-160 ft.). Given that the compaction ratio is from 2 to 4, this means that the peat had to have had a mean thickness of from 61-122 m or from 200-400 ft. This is quite a lot of peat development, divided into 27 layers.

Significantly enough, the writers here make no mention, neither above nor below, nor in-between any of the layers, of the fossil soils that would have been necessary for the growth of all that peat, and certainly not the supposedly necessary 27 soil layers in this case. The 27 lignite seams clearly imply that the basin rose and fell 27 times over, and that is just not believable. And where, by the way, are all the actual *swamp*-type fossils that should reasonably be expected to be found in what, according to these orthodox scientists, is essentially a fossilized swamp, or 27 of them, stacked one above the other?

The entire lignite series is itself overlain by supposedly Pleistocene sediments such as sands, marls, clays, etc., which contain fossils and "coalified" plant remains. Above these are fluvial sediments consisting of coarse to medium sands and conglomerates of uneven thickness and which are discontinuous over the basin. These, in turn, are overlain by more fluvial and supposedly limnic deposits consisting of sands, clays, marls, and conglomerates, again discontinuous.

The total thickness of these top, recent, sediments is from 15 m (50 ft.) in the east to 150 m (500 ft.) in the west, which seems like an awful lot of material for such recent deposition. And far too much for any uniformitarian process, regardless of which one was ongoing. And by what agent were conglomerates, of all things, delivered into an elevated basin?

From their analysis of the plant matter in the lignite, the writers deduce that the plants grew in a forest with occasional changes to an open water environment and from a swamp to a fen, with changes also from still water to flowing. Therefore, to explain the lignite we have need of forests, lakes, rivers, fens and swamps, alternating somehow or another, and interrupted regularly by the deposition of clastic sediments, clays and marls. They do not say what kind of plants are involved, or how they could have got from the forest into the swamp, a swamp that was extensive enough to occupy a basin up to 20 km (12.5 m) across.

How all this could come about so as to produce 27 layers of lignite that are thick in the west and thin out (wedge out) to the east, is beyond me (uniformitarian-wise, that is). The presence of the layers of sediments in between the lignite layers, instead, suggests to me that the entire sequence, from bottom to top, i.e., from the basal conglomerates to the top of the Pleistocene, including the lignites, is a straightforward series of sedimentary deposits filling the basin by means of repeated back-and-forth flooding during a major flooding "event," or a number of such "events."

When we look at other lignite deposits in the area, e.g., the Anchlada mine, in the same basin complex, we get a similar story, but the authors here (Oikonomopoulos et al., 2008) give us a bit more detail about the sub-basin in which the lignite lies. At the base, and immediately overlying the bedrock are conglomerates, marls, sands, and clays below the lignite. No fossils, or what may be termed "relict" soil is to be found.

There are a number of lignite beds and these alternate with clays, marls, sandy marls, and sands in the lower levels, while in the upper levels, we find fluvio-terrestrial conglomerates, lateral fans and alluvial deposits, all of these representing, as we know, a lot of water moving quickly. The fact that the series begins (as usual) with conglomerate directly on the bedrock implies that the bedrock was swept clean or scoured out prior to that conglomerate's deposition, it being a "basal conglomerate."

The authors have determined, from plant analysis, that the environment was: ". . . open water, reedmoor, Taxodiaceae (bald cypress, sequoia) forest, and mixed forest environments," which is quite a few different environments. It would seem, therefore, that they found a lot of woody material in the lignite. They consider that the basin was once a "floodplain environment which included a large meandering river system." This, of course, is the tiresomely standard explanation for every basin.

It is so because this scenario is the only one by which a basin can, theoretically, be filled up in an approved uniformitarian way, as we've discussed earlier, i.e., the river meandering across the valley. Supposedly, the basin was "little by little" (what else?) converted into a wet forest swamp, such as are common enough, with clusters of trees (the bald cypress no doubt) which contributed to peat formation, like everything else.

After this, yet another "flood episode" stopped the peat formation and deposited a thick clayey layer. As to how a basin could "little by little" be converted to a "wet forest swamp" the authors do not say, nor do they give any evidence of this conversion process and its supposedly gradual development (Oikonomopoulos et al., 2008). One presumes another leap of faith is required here.

In the profile shown as fig. 2 in their paper, the authors define two major lignite series, which they describe as xylite-rich for the lower, with no evidence of plant roots, and the second as xylite-rich, interspersed with matrix-rich layers, with the xylite-rich layers dominating towards the top of the series. *Xylite* means wood-derived, or woody, and is associated with the break-down of woody material, as in the formation of the solvent xylol (or xylene), wood alcohol.

Here, geologically, it relates to the composition of the lignite as being woody and the authors tell us that: "The xylite texture of the deposits is amplified by the occurrence of large and small xylite parts as branches and stems. Roots are also observed, . . . and deposition continued inside the lignite horizons. Moreover, two smaller stems are depicted and another stem with well-preserved tissue is seen in a vertical position in a huge inorganic clayey bed . . . this particular stem continues into the overlying lignite horizon, suggesting an autochthonous (grew in place) character."

These "stems" by the way, are just a poor translation of large "tree trunks," as we'll soon see. The "roots" that were observed are actually the root-balls of trees and do not imply that the trees grew in place, as they would remain attached to a floating tree.

The authors deduce from various analyses that the "peat" accumulated under "highly saturated" and "anoxic conditions," while the plant samples point to a forest vegetation; this, to me, is something of a contradiction, since *sequoia* does not grow in swamps and even bald cypress is more at home in flood plains and river channels, where water movement tends to oxygenate the environment. The good preservation of the material identifies it as deriving from lignin-rich plants, especially woody species, and implies, therefore, that these lignites are, for the most part, made up of woody material.

This is surprising, since many of these trees do not grow in swamps, and peat is supposed to be formed from swamp plants, and there is a very large amount of lignite in these Greek deposits. This is the sole reason for invoking a "wet forest swamp" which apparently was periodically converted into a limnic one (lakes and ponds) neither of which, as noted, would support *sequoia,* or most of the other tree types.

The presence of many discrete, independent xylite bodies (tree trunks and branches) in the lignite cause the authors to suggest piedmont plains, badly drained floodplains, clusters of trees, streams bringing material into the floodplain from surrounding hilly areas, and inundation (flooding again) converting the floodplain to a wet forest swamp, and then a reedmoor vegetation grew, along with various types of forest trees. Such constantly ongoing and implied gradual change could almost be called, oxymoronically, "uniformitarian changeability."

Obviously, explaining the origin of the lignite with respect to all the different types of plants found within it is difficult in light of the fact that so many of these plants do not grow in swamps—and swamps are the only allowed origin of peat bogs. Such plants are sequoia, maple, basswood (linden), bayberry (wax myrtle), birch, beech, pine, palm, magnolia, daisy, alder, and elm, among many other types, only some of which are found in wetlands, such as sedges and cypress, and hence, the latter two may be the source of the reedmoor suggestion. The others only grow on dry land, well away from swamps.

Of interest is that one of the tree species encountered was a sequoia, which the writers consider "remarkable" but don't say why; it's no more remarkable than some of the others, although, along with the bald cypress, it does tell us that Greece, in the not too distant past, had trees very similar to the present-day Americas, suggesting some land connectivity at some point to explain their spread. The authors also report the presence of lianas and fungi, which have no love of swamps and only grow in drier and warm conditions. The fungi here are not swamp fungi; these are forest fungi.

Also present are swamp cypress, which imply a warm and wet climate, so the plants found here either indicate a number of changes in climate, or the plants were transported into the basin from many different places (Oikonomopoulos et al., 2008).

Therefore, the depositional multiplicity of environments of this basin are considered to have been: ". . . a floodplain . . . in combination with the formation of oxbow lake [sic] within the abandoned channel, represents a landscape with predominantly swampy conditions and 'periodical flood episodes' together with the deposition of fine sediments." Floods are yet again called upon to explain the evidence, but the conglomerate, sand, and clay layers in these lignite deposits are all of the order of 1-10 m (3-30 ft.) in thickness, representing floods of a far greater magnitude than any seen anywhere today.

Because they found an upright tree trunk, which they tentatively identified as growing in place, extending from the lignite up into the overlying clay, the authors are obliged to invoke quick burial by a "flood episode" to account for the trunk not having rotted away. This is a good example of the "polystrate tree" problem (*polystrate* means "many strata") which has been a perennial thorn in the side of uniformitarian geology for almost 200 years. Fig. 8-2 shows a typical example.

Fig. 8-2: Example of a polystrate tree, a favorite of creationists. The hammer is about 15 ins. (38 cm). Image: CC-BY-SA-3.0 license. Photo: Michael C. Rygel (cropped) Wikimedia Commons.

A polystrate tree is one that is in an upright and presumed growing position, but penetrates through two or more strata, as shown in fig. 8-2, strata that are naturally supposed to have been deposited over long ages. Such tree trunks have even been found in coal mines penetrating though multiple seams. Obviously, the problem is that the trees haven't rotted away, and hence must have been buried suddenly and completely. Thus, all the strata they penetrate must have been laid down one after the other in very short order. And by that, I mean only hours or days, especially hours.

Following up on Oikonomopoulos et al. (2008), with respect to finding trunks of trees in the Achlada lignite mine, Iamandei et al. (2012) tell us that "huge" trunks were discovered, one such being 6 m (20 ft.) tall, which was removed and placed in front of the mine's administrative building. This tree was identified as *Taxodioxylon taxodii*, a swamp cypress, very similar to one commonly found in the southern U.S., Mexico, and Guatemala at present.

Their description of the mine-deposits follows Oikonomopoulos et al. (2008) but they also mention finding fossils of rodents and lago-morphs (rabbits or hares), which, of course, do not frequent swamps.

A number of trunks and other plant stems were found in what was interpreted as the growing-position, since they were vertical, and they extended through the overlying thick muddy clay layer and up into the next lignite horizon. We recall, of course, from our dis-cussion of the trees in the Columbia River in volume 1, that vertical tree trunks do not necessarily mean growing-position, because they may have been floated vertically in a deep flood, the heavy root-ball keeping them vertical.

While the writers here rather casually pass over this "thick, muddy clay layer" as though it were of no import, the reality is that a normal, average flood does not deposit a thick layer of clay, or anything else. Any normal flood deposits no more than a few inches or cms of mud, at the most, and only in some places, while usually it's no more than fractions of an inch or centimeter, in general. Therefore, the flood required here, as in every case we've exam-ined, must have been inordinately large, deep, and widespread, as is implied by the thick layers of conglomerates mentioned repeat-edly earlier.

Most flood deposition actually occurs in the estuary or delta of the flooded river, as the normal result of the river meeting the sea. In the floodplain, deposition is selective and occurs at certain places where the flood is embayed or obstructed, as we see in towns, in particular.

Fig. 8-3: Tree trunk on display outside the Achlada mine administration building, on the left, with a close-up showing the mummified nature of the wood on the right. Image: Iamandei et al. (2012).

The trunk placed in front of the mine building was originally 10 m (33 ft) tall and 2 m (6 ft) in diameter; the upper 6 m (20 ft) section was removed and the remaining 4 m (13 ft) section left in place. The authors describe the wood as being more mummified (dried out) than coalified (Iamandei et al., 2012), which means that I might (heretically) suggest that it hasn't been there all that long, maybe some thousands of years.

While the lignites are generally explained by referring their origin to swamps and mires such as we see today, the presence of vast amounts of xylite, wood fragments, branches, twigs, and even trunk-sections would tend to contradict that explanation, especially as most of the trees and other plant types do not grow in or any-where near swamps. As we will see in a later volume, the same can be said of the plant material we find in coal seams. There, as noted, we sometimes find trees extending through multiple seams and the intervening shale/limestone layers, and we also find boulders of considerable size.

Fig. 8-4: Just to show what a lignite mine looks like. Ptolemeida open pit lignite mine Macedonia, Greece. Image: ID 247399709 © Vasilis Ververidis, Dreamstime.com Royalty-free License.

As far as the actual composition of the lignite goes, macro- and microscopic analysis shows it to be composed mostly of small fragments of wood, in the nature of wood chips, like, for example, the coarse sawdust produced by a chain saw or a wood chipper. A study done for the U.S. Dept. of the Interior on Greek lignites examined the fine material and classed it as xyloid, or woody material, and described it as attritus, which is defined as pulverized or finely divided material, often specifically referring to plant matter. The study further says that the lignite: ". . . superficially resembled large chip-like fragments of wood" (Toenges et al. 1951). Because the *xylitic* fragments and attritus could only derive from actual wood, a "superficial" resemblance to wood fragments is to be expected, I would think.

Fig. 8-5: Face of lignite mine showing lateral extent and parallelism of the beds and perfect evenness of the layers over considerable distances, as though they were laid down in sheets. Image: M. Galetakis.

The two photographs (fig. 8-5, 8-6) show the typically layered lignite deposits of Greece. The flat, even layers with sharp layer-to-layer contacts are easily seen. No unevenness as might be expected from a bog or lake and no sign of fossil soils are apparent, especially at the contacts between the pale-colored layers (marls) and the overlying lignite seams. These contacts are sharp, with no transition between the layers, and the extremely thin layers are a particular puzzle, especially where we see alternating thin layers of different material, all perfectly flat, even, and continuous.

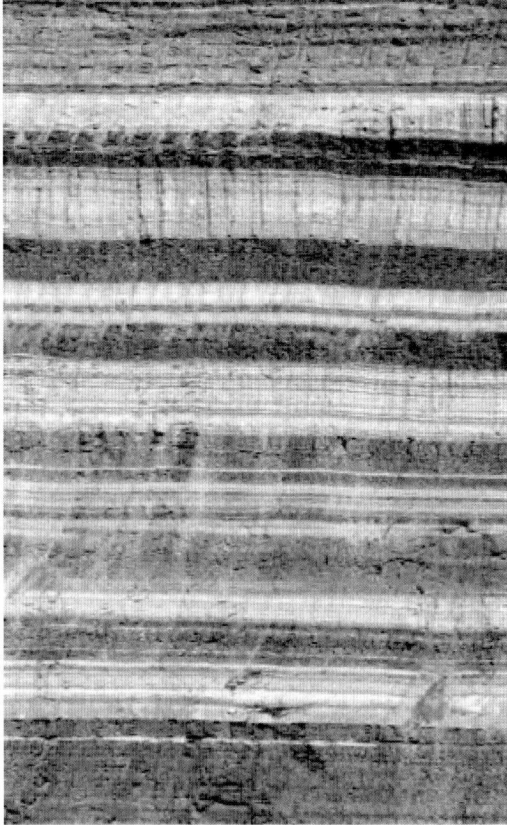

Fig. 8-6: Close-up of lignite layers, again showing generally even and persistent thickness of the layers, with only a hint of undulation. Image: Michalakopoulos et al., 2014.

Note the sharp contacts between layers of different composition. Sharp contacts imply sudden changes of conditions. The multitude of such layers signifies, according to standard theory, an equal number of changes in depositional environment. Here, the dark bands are obviously lignite, while the pale bands are without doubt marl, a fine limestone typically found in lakes. The few tan/brown layers are likely sand and/or clay but I can't identify any soil layers, the lignite appearing in most cases directly underlain by marl.

In both photographs, the marl layers are exceedingly flat-surfaced and of remarkably even thickness for old lake beds, in this case about as flat as a billiard table. Note the extremely thin layers which persist perfectly evenly and with no break across the full extent of the mine face in fig. 8-5. All of this occurred slowly over eons of time, in exactly the same way, at exactly the same rate, everywhere? I've my doubts.

In fig. 8-5, I count 25 lignite layers, separated mostly by what appear to be marls and clays, which, according to convention, implies 25 changes of "facies" from peat bog/swamp to lake and back again, each one more or less perfectly even, many extremely thin and all with very sharp contacts above and below. There is no hint of the hummocky unevenness typical of any bog or swamp to be seen "in action" today, where clumps of grasses, plants, bushes, or trees are interspersed with pools or sluggish streams or sheets of slow-moving water, as might be seen in any large swamp such as the Florida Everglades, Louisiana Bayou, or the Pripet Marshes in Poland/Belarus/Ukraine.

The following series of photos show some well known and very extensive present-day swamps, occupying vast areas of exceedingly flat territory. As our own eyesight and common sense tell us, it wouldn't matter how long one waited, none of these swamps is ever going to fill up to the point of producing a perfectly flat surface such as we see in the lignite mines of Greece, or anywhere else. This is most obvious in the case of the Pripet Marshes, Belarus.

In the Drama basin, again in the north, A. Georgakopoulos (2000) informs us that the Neogene-Quaternary formations consist of clay, mud, sand, conglomerate, marl, peat, and lignite, which is to say, it's just like everywhere else, with the lignite extending all over the central plain of the basin. The lignite here is considered to be only 1 my old, which is very young compared to the lignites in adjacent basins. This does not make a lot of sense, considering they're all part of the same basin complex and presumably share a similar tectonic history.

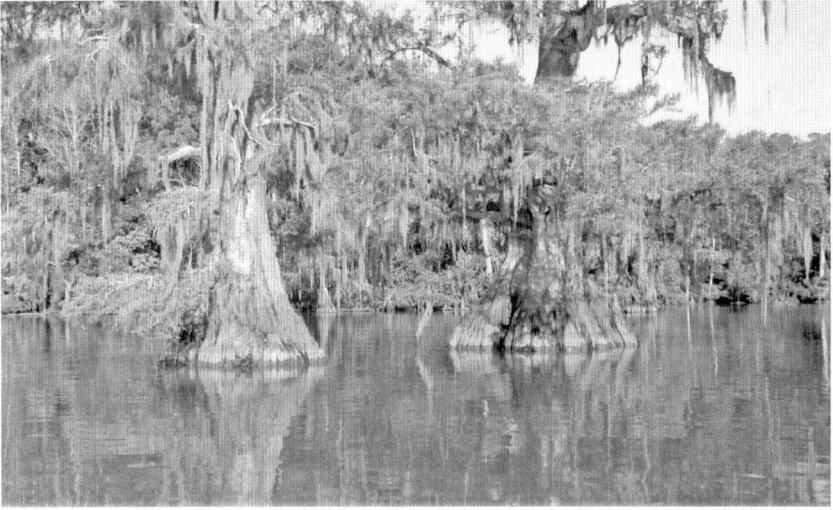

Fig. 8-7: Louisiana Bayou, Louisiana U.S. showing land and trees in water and how uneven the surface of the bed of that swamp must be. Image: ID 52266638 © Joe Sohn, Dreamstime.com, Royalty-free Lic.

Fig. 8-8: Pripet Marshes, Belarus. While generally flat, the unevenness is obvious, meanders are everywhere, but no peat bogs are to be seen. Image: ID 218149458 © Viktarm, Dreamstime.com, Royalty-free Lic.

Fig. 8-9: Florida Everglades. The swamp is obviously uneven with land and water, and if it were drained would not produce the perfectly flat surfaces we see in the lignite mines of Greece. Image: ID 83644575 © Romrodinka, Dreamstime.com, Royalty-free Lic.

In fact, all the basins of Greece are supposed to have shared a similar tectonic history, since such tectonism is regionally acting, and individual basins couldn't have experienced anything other than minor, narrowly focused forces, and movements, related in a secondary way to the main movements. It is, therefore, clear that all the basins do share a common history, and applying different ages is, again, I suspect, merely an effort to divide up what is obviously a singular major event into a number of minor events spread out over time, in the complete absence of evidence.

Clastic material from alluvial fans was deposited at different times and divides the lignite deposits into three units, or sequences, of lignite beds interspersed with muddy marls (gyttja) and sands/silts, much like the others we've seen. While *gyttja* means muddy, it does not mean soil. While the author says "clastic material from alluvial fans," it's reasonable here to ask where all those "alluvial fans" came from, since, as we've seen earlier, they typically derive from floods out of mountain valleys, and here again the authors reference the usual three units. There is, as we know, an inherent contradiction between the quiet environment of a lake and the floods required for alluvial fans.

In another pertinent paper from Metaxas et al. (2010) on which Georgakopoulos was one of the collaborators, we are told that, beginning about 5 my ago, the basal sediments in the Ptolemais basin came from the north and west, supposedly because the Florina basin in the north lifted up, which it would have to do, otherwise any material coming from the north (in a uniformitarian fashion) would be trapped in it. We're also told that the Florina and the western margins of the Ptolemais contain much coarser sediments (coarse sands and conglomerates) than the eastern margins, which contains silts and clays.

Therefore, whatever deposited all the sediments came from the north and west, and whatever it was, while being capable of transporting all kinds of coarse and fine materials, deposited all of its heavier constituents in the north and west and then moved south and east to deposit all of its finer material, clearly implying a major flood, like everywhere else.

As can be seen in the hill-shaded relief map of fig. 8-10, the Ptolemais is separated from the Florina by an unmissably distinct ridge connecting Vernon Mt. and Vorras Mt., while the Ptolemais graben is bounded on the west by Aksion Mt. Apparently, therefore, whatever brought in and deposited those sediments came from the northwest, swept through the Florina basin, then up and over the intervening ridge and filled the Ptolemais basin, just as we saw with the Amorgos/Astypalea basins in the Aegean. I'll wager there were no ups-and-downs involved here, and the Florina didn't move.

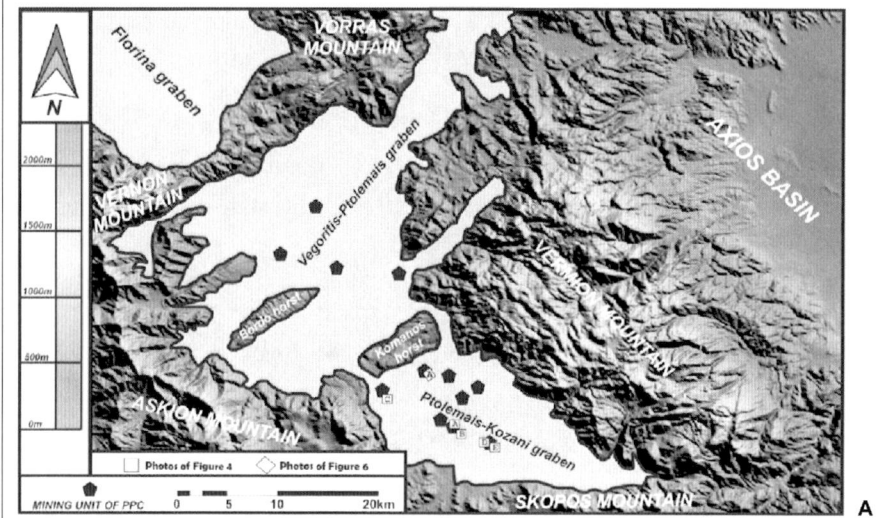

Fig. 8-10: Relief map showing ridge separating the Florina from the Ptolemais, Askion Mt. bounding the Ptolemais to the west, as well as the Borrio and Komanos horsts presenting further obstacles to sedimentary transport from the north and west, leaving us wondering how the sediments filled those basins, and reminding us of the Aegean. Image: Diamantopoulos, A., 2006.

The authors, like many others before them, invoke "quick flow" and "great volume of material" to account for the evidence, and then, since lignite cannot form under such conditions, declare that later, when the lignite formed elsewhere in the basin, sediment came from the east. Tectonic movement supposedly caused the drainage to change from the north and west to the east (quite conveniently). The central and eastern part of the basin then supposedly subsided so that lignite could now form in the Ptolemais.

It is, of course, clear that all these proposed tectonic movements are mere contrivances devised to enable a uniformitarian-sounding explanation to be applied, and this would, in fact, appear to be another "exercise of the imagination."

The big problem here is accounting for the material from the north, the rock type being distinctly different, and only occurring to the north and west, reminding us of the situation in the Voidomatis valley earlier, where we also found "foreign" material. Further, the authors tell us that immediately prior to recent times, i.e., the end of the Ice Age, the basin was covered by a layer of conglomerates with interbedded sands, clays, and loam, a terrestrial fluvial series.

So here, therefore, these Greek "orthodox" scientists are plainly stating that this basin complex, extensive as it is, was overrun by a major flood at the end of the Ice Age. We have seen enough by now to recognize the meaning of "covered by a layer of conglomerates with interbedded sands . . ." Interbedded sands and conglomerates obviously imply that a series of floods, one after the other, or the same one ebbing-and-flowing, swept over this basin at the end of the Ice Age.

Total sediment thickness in the Ptolemais basin is given as 500-600 m (1,600-2,000 ft.) and up to 1,000 m (3,300 ft.) in places (Delogkos, et al., 2018) and, as can be seen in fig. 8-10, more or less completely fills the entire multi-basin complex. Again, we are reminded of all those basins and coastal plains of the Aegean area that were also filled up with "coarse, clastic sediments" that came from the north or northwest.

Metaxas et al. (2010) invoked some complex tectonic movements to account for the infill of this basin complex and the creation of the (assumed) conditions for the growth of peat, and the development of lignite. These movements, extending over a long period needless to say, supposedly involved rising, subsiding, and tilting of both the grabens and the surrounding mountains and horsts, and two (fig. 8-10) have been identified within the Ptolemais basin.

However, Delogkos et al. (2018), in an analysis of the structural geology of the lignite mines in the Ptolemais basin, only detected two extensional episodes. These were related by Delogkos to movement on the faults bounding the graben. Such faults are oriented in the familiar NW-SE and NE-SW directions seen in the Aegean area, indicating stretching in these directions.

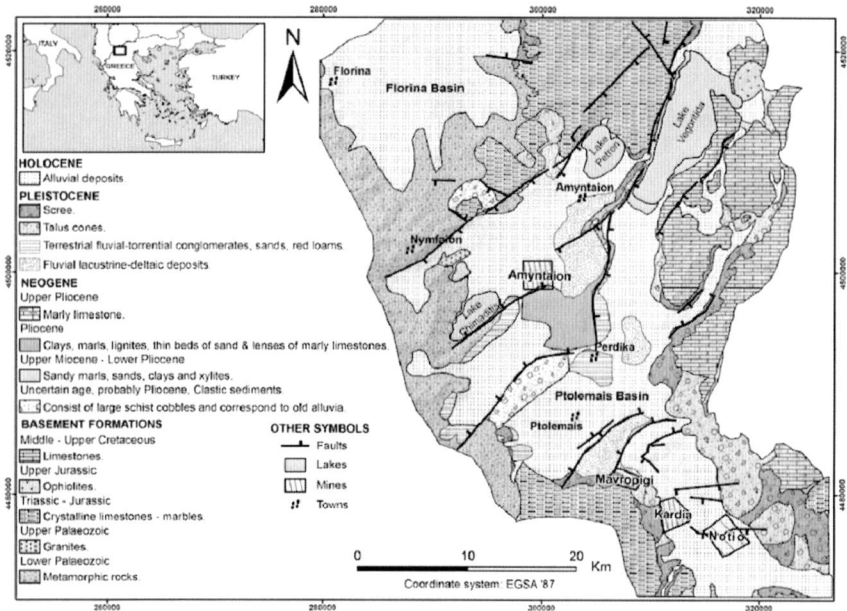

Fig. 8-11: Geological map of the Florina-Ptolemais basin complex. Note that the Pleistocene sediments contain "fluvio-torrential" conglomerates. Image: Delogkos et al., 2018.

The first movements pertained to the formation of the graben, affecting only the basement rocks, while the second phase, or event, came after the filling of the basin and the formation of the lignite deposits, since, according to Delogkos et al., the fault zones, or arrays, affect the full, or at least the exposed, thickness of the lignite deposits. At the same time, these authors suggest activation, quietude, and reactivation of these faults over long time periods rather than the single "event." This, of course, is necessary to get all that repeated up and down movement.

Fig. 8-12: Faults in lignites, Ptolemais Basin, N. Greece. Shown here is the upper terrace of the multi-terrace arrangement of a typical strip-mine. Faults extend from the base to the top of the lignite and no disturbance is seen. The top of the series has been planed-off by some erosional agent prior to deposition of the thin top, most-recent layer, which looks to be about 2 m (5-6 ft) based on the individual at the bottom. Image: John Walsh, Fault Analysis Group, UCD, Dublin, Ireland.

The fact that these faults extend from the bottom to the top of the deposits, as shown in fig. 8-12, and with little sign of disturbance along the fault surfaces, suggests to me that they're of recent formation, and my own opinion is that the basin formed during one disturbance, filled up, and suffered a second, recent disturbance, in parallel with other recent disturbances, such as in the Aegean. If these faults had been repeatedly activated, there should be crumpling and distinct evidence of repeated disturbance along the fault surfaces, which there clearly isn't, as the fault surfaces are neat, as can be seen in fig. 8-12.

Moving to the south of Greece, we go to the Megalopolis basin which has lignite seams interspersed with the usual layers of marls, clays, etc. While this basin has not been as intensely studied as those in the north, Karkanas et al. (2018) tell us that at the base of the sequence lies a series of bedded sands/silts characterized by ". . . relatively high rate subaqueous sedimentation . . . with local evidence of slumping and liquefaction attributed to a seismic event." Further, they report that the upper part of the sequence, curiously ". . . follows a major hiatus attributed to exposure and erosion. A series of erosional bounded depositional units are observed in this sequence suggesting important water level fluctuations."

Thus, what we have here are a number of depositional units each laid down on an eroded surface and each with its top surface eroded in turn. We will, of course, immediately recognize this as a classic multiple flood profile, such as the flood sequences we discussed earlier, and as can be envisaged in the following scenario:

A flood deposits a unit of coarse to fine material grading upwards. The next flood comes along, erodes the top of the last one, deposits another unit on the erosion surface, and repeats the process any number of times; the lignite, being the lightest, will be last member of each unit to be deposited. As a matter of fact, and not a little surprise, this is just what Karkanas et al. (2018) propose occurred, though not by any means in the short time frame I might have in mind.

According to these writers: "The sub-aqueous emplacement of the deposits is attributed to sub-aerial flood-generated, organic- and carbonate-rich dilute mud-flows and hyperconcentrated flows. Organic-rich sedimentation culminates with the formation of the overlying lignite seam." Here the authors suggest a transport and sedimentary origin for the lignites (not a "floating mat" mind you), but it's a different proposition altogether to the explanations proffered for the northern lignites.

These writers are, needless to say, interpreting these changes as occurring slowly: ". . . a gradual return to warm and humid conditions of an interglacial period . . ." though I can't really see how a sequence of floods implies "gradualness." Near the top of the sequence is an erosional surface upon which are found archaeological remains, and some fossils, all buried by recent mudflows that also redistributed them.

Thus, sometime in the very recent past "mud-flows" (which might also be called what they are: mud-laden floods) eroded the (loose) surface of the ground, picked up various archaeological remains and fossils, and redistributed them, which is to say, scattered them around, as floods do.

Athanassiou et al. (2018) report that an extensive alluvial fan covers the sequence in the central section of the basin near the city of Megalopolis. A survey conducted by Thompson et al. (2018) reported that the basin switched back to a fluvial one later in the Mid-Pleistocene and they report the presence of ". . . layers of clayey silts and sands and . . . thick gravel layers." Above these lie terraces, clays, silts, sands, loose conglomerates, and scree. Finally, there are alluvial fans along the western margin of the basin that lie at high elevations, referred to by Thompson et al. (2018) as: ". . . notably above the basin."

This is "notable" because these fans are supposedly older than all the other Pleistocene deposits, dating, supposedly at least, from the early Pleistocene. Whenever they actually date from, the fact that they comprise alluvial fans and gravels against the western side of the basin, and so high up as to be "notably above the basin," suggests that a lot of water must have come out of, or more likely over, those hills, from the west.

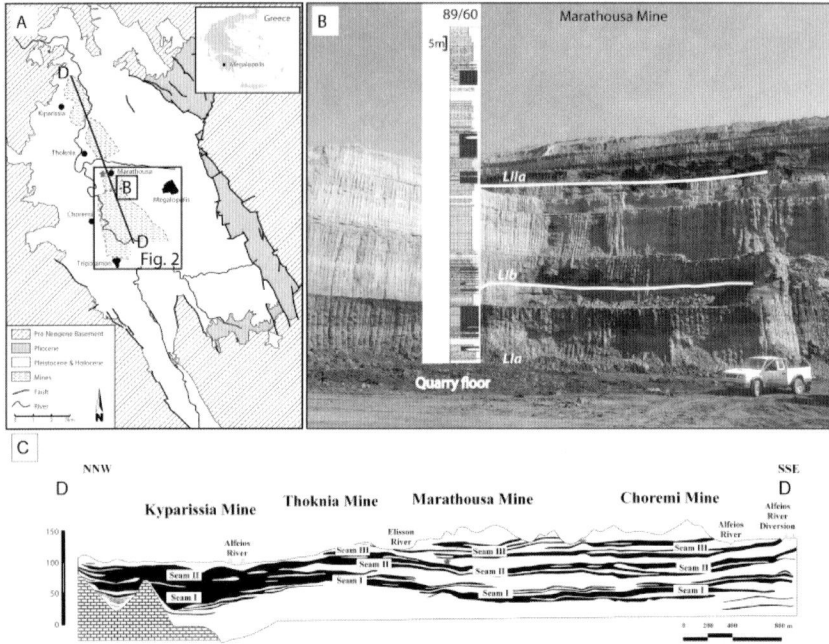

Fig. 8-13: Map and image of the Megalopolis Basin with NNW-SSE section D-D through the basin infill showing how the lignite seams merge to the northwest. It has long been a puzzle as to why coal seams merge and diverge as shown here. The vertical exaggeration of the section is at least 3:1 Image: Tourloukis et al. (2018).

While explaining the elevated alluvial fans is problematic, we must not miss other no less problematic phenomena in this basin. Thompson et al. also mention "terraces" above the present surface of the basin. Terraces, as we know, are put in place by flowing water, as in any river or flood. Therefore, apart from water coming over the hills, we also need large quantities of deep water flowing through this basin to emplace those terraces. As can be seen on map A of fig.8-13, the basin is oriented to the northwest, the typical direction whence so much of the sediments on Greece derived, with major floods being the only possible agency. As the section D-D of fig. 8-13 shows, this also appears to be the direction from which the lignites came.

Significantly, at least one team of distinctly orthodox geologists are prepared to allow a sedimentary origin for some lignites in Greece, though the evidence gave little choice in the matter. The lignites at Megalopolis appear to be similar to those seen elsewhere, being mostly made up of pulverized woody material, and this suggests to me that they all have the same sedimentary origin. This would explain why the theories of formation proffered by most workers don't explain those lignites in any kind of realistic and reasonable way.

The effort to explain them is, as I've said, predicated on a required, and strict, adherence to uniformitarian geology, and hence all workers can only invoke something that goes on at present, and has some relation to swamps, mires, bogs, or waterlogged floodplains, along with the tectonic movements we've all been persuaded are constantly going on.

A depositional origin of peat, however, would provide us with an explanation for those mysterious sapropel layers found in the Aegean, and also explain why no lignite was found there. It seems to me that the sapropel layers are, in fact, simply the undersea equivalents of the terrestrial lignite beds, and I aver this to be the case.

Thus, no mysterious, and highly unlikely "anoxic event" is required for their explanation; they are simply layers of organic materials deposited in Aegean basins by one or more "flood-events," as have been repeatedly invoked in Greece by our uniformitarian colleagues. The sapropels were quickly buried by more sediments, sealing them off from oxygen and initiating anaerobic decomposition, which is, I'm suggesting, little different from what occurred in the lignite basins.

At this point, we have looked at enough of the lignite basins in Greece to get a general impression of their makeup. All, including ones not mentioned herein, display much the same features, sedimentary sequences, and composition. Only in one area is anything different reported—Crete—and that report comes from Karageorgiou et al. (2010).

Crete has a number of lignite deposits, and the sediments in the basins are much the same as everywhere else, conglomerates at the base from cms to tens of meters thick, lying on a typical erosional bedrock surface, (another basal conglomerate) and overlain by various other sediments interspersed with lignite seams.

Of interest is that the authors tell us that they found evidence indicating that: "Frequent sea transgressions are also observed separating the lignite beds." A "sea transgression" means that the sea invaded the land, and the usual explanation is that either the land sank or sea level rose, always slowly, of course. "Frequent sea transgressions," however, is a term we don't hear very often.

In this case, given that sea transgressions are reported only from Crete, rising and falling sea levels in the normal long-term Ice Age to Interglacial cannot be the answer because Crete and its basins are well above sea level. The only other (orthodox) option would be the same as every other basin, i.e., that Crete itself rose and fell "frequently" but, of course, slowly, except that Crete is part of the European plate and is not tectonically independent; the Cretan Trough seen in fig. 2-1, is only a surface crack, albeit big. At the same time, Africa is supposed to be pushing up under Crete and therefore should be lifting it up, rather than allowing it to subside.

When describing the coarse sediments in one area, the authors describe them as "fluvial-torrential" or "marine formations." The latter consists of a "transgression conglomerate," which certainly implies the rapid movement of lots of water, and the presence of oyster shells and other marine fossils such as *neritina*, the tiger snail we met earlier 600 m up in the mountains, clearly implies that the water came in from the sea, and we've already seen lots of water effects around Crete's shores.

As can be imagined, it takes a certain effort to sweep up stones and debris, and large sea-shells, and carry them in over the land and up into an elevated basin. Thus, it would seem that the evidence from Crete suggests that the sea invaded the land in the rather violent manner of a major flood event, which was repeated multiple times. The oysters, and other shells, on Crete are high up in a cliff near Matala, and are mixed with sand and gravel. And much of the foregoing, as we recall, echoes Spratt (1865) and others.

While no lignite is reported from the basins of the Aegean, lignite has been found on at least some of the islands, including Rhodes, Kos, Samos, and Alonissos and in the usual kinds of basins. Taking the latter island as representative, Mantzouka et al. (2019) report that the deposits are much the same as elsewhere, the usual conglomerates, marls, lignites, and clastics, etc. Here, however, we also have red clays and "continental conglomerates" (presumably as opposed to "marine") overlying the lignite-bearing sequences.

Considering that such lignite basins are found all over Greece and on islands, it's worth asking if they're found anywhere else in the region. And they are; in fact, they're found all over the greater southern European region, and across into Türkiye and up into southern Russia. Albania, Italy, the Balkans, Bulgaria, Hungary, Poland, Central Europe, Ukraine, Belarus and Russia, among many others, all contain lignite deposits much the same as the Greek ones, and all consisting of interstratified deposits of conglomerates, sands, clays, marls, and lignites, all of approximately the same orthodox age, according to various writers (whether or not that is the actual age of these deposits).

Whatever the case may be, we know what conglomerates, gravels, and sands imply. The fact that the very same phenomena are to be found over such an extensive area suggests to me (and Occam) that a large-scale, super-regionally-operating agent was responsible, rather than the extremely unlikely scenario that the same thing happened at multiple times during the same extensive time period in any number (hundreds) of individual localities, at different elevations, and spread over the vast area that is southeastern Europe, Ukraine, southern Russia, Türkiye, and on up into northwestern Europe.

Continuing with the basins of the Aegean, we have reports of sapropels in various basins other than those identified by Isler et al. (2015) earlier. A second paper from Isler et al. (2016) the following year tells us that sapropel deposition is (naturally enough) related to an increased supply of organic sediment into the sea. The question is why that organic material wasn't eaten or oxidized, and the authors have no explanation, as do none of their peers.

However, they do provide a date for the last, topmost, sapropel layer, dating its base to 9,900 years ago, which is to say about 10,000 years ago (give or take a couple of thousand). The same age was provided by Triantaphyllou et al. (2015) for sapropels from the North Eastern Aegean, 10.2-8.0 thousand years ago. There is no question but the organic material derived from land, and I conclude that it is simply the marine equivalent of the lignites, deposited by the same "event."

Given what I consider the inadequacy of orthodox explanations for much, if not all, of the evidence found regarding the lignite deposits of Greece, and not to mention all the other phenomena we have discussed in the last few chapters, I consider that the condition of Greece and the Aegean, as we now find them, is due to entirely different causes than those invoked by uniformitarian geology.

The agents and forces necessary to accomplish the work we've seen done need be of a far greater magnitude and operating on a far larger scale altogether than any that we see to be in action in the present day, or even during an occasional so-called catastrophe such as we're used to. All the studies and reports we've looked at, nearly to a one, contain references to floods and mega-floods, depositional events, seismic events, paroxysms, etc., while the word "diluvial" is also thrown in here and there.

Something else that stands out and that I've mentioned a number of times is that in all their descriptions of different geological sites and environments, not a single author gives any evidence that points to the action of everyday processes going on for long ages. The usual millions of years should be represented by millions of thin, yearly flood layers, or at least some portion of millions if we are to accept that everyday causes were in action in any way effectively, as they're supposed to be.

Collectively, the authors instead infer by their own language and descriptions, and even stated interpretations, that, on certain discrete occasions, over the course of long ages totaling about 20 my, there occurred a number of "events" of various kinds, such events typically involving flooding and tectonic uplift. Events are, of course, and by definition, short-term occurrences.

Despite the wishful thinking and allusions of the uniformitarians, an "event" does not go on for any length of time; it is measured in minutes, hours, or days, or at the very most, a week or so. The aftermath may certainly linger for a while, and then just fade away, like an old soldier as the saying goes, but this does not take very long, and certainly not millions of years.

In geology, it is only recently that the word *event* was even allowed into the lexicon, and this by forced necessity, and with maximum pressure applied to restrict its use to such occurrences as would be considered just somewhat out of the ordinary, i.e., extra-large catastrophes such as we're used to, such as, e.g., floods and tsunamis, the only watery ones allowed, and, of course, underwater floods known as turbidity currents.

Aside from these occasional "minor" events, the world is officially considered to have continued on its calm uniformitarian way, as it has always done. In which case, the signatures, or evidences, of all these "events," defined as, and by, discrete sedimentary units or sequences, such as those seen on the terraces of the Voidomatis River, or thick beds of conglomerates, or giant alluvial fans or so-called "alluviation" events, should be interspersed with long sequences of yearly deposits.

And these latter should be traceable to the everyday actions of whatever standard uniformitarian processes were considered to have been ongoing at any particular place, over the course of all the other time not occupied by the few short-term occurrences represented, as we've seen, by those out-of-the-ordinary depositional series found on Greece.

The most usual agent called on to perform geological work is the river, derived of course from rainfall, and this agent is most popular because it operates almost everywhere except deserts (though even here it can operate in the form of occasional flash floods through wadis). The second most popular agent is ice but that is limited to the polar regions and high mountains.

Rain and rivers, on the other hand are almost universal and are credited with an amazing amount of work, cutting out huge valleys, laying down vast sedimentary deposits, wearing away the hills and landscape, and grinding it all down flat, until there's nothing left but an immense "peneplain" on a level with the surface of the ocean.

We've already discussed the validity, or otherwise, of the "venerable" peneplain concept. We've also discussed the ability, or inability unless it's carrying significant amounts of debris and sand, etc., of rivers to erode hard rock, so we are not at all surprised when geologists are forced by the evidence to invoke agents far more powerful than any ordinary river, whether it's in flood or not.

When the writers do refer to a "flooding event," they correlate it with the "end of the Ice Age" or some "change of climate," each of which, as their thinking seems to be expressed in their papers, took thousands or tens of thousands of years. Their implication is that these so-called events, all represented by thick beds of conglomerates, boulders, masses of coarse sand, thick beds of woody lignite, etc., are the result of slowly acting processes. The "event" is actually to be thought of as nothing much out of the ordinary, stretched out, as it implicitly was, over a long time period, as uniformitarianism dictates.

However, the writers' phraseology, and the evidence they describe and seek to explain, clearly indicate short-term, highly energetic events, as I have pointed out. They make no mention of any evidence indicating any long period of attenuation. Instead, they seem to depend on some sort of leap of faith on the part of the reader, based on the assumption of a mere *belief* in uniformitarianism.

Not only do they not describe any definitive effects or results of the normal everyday processes, or give any examples, but they invoke any number of purely *theoretical* agents and processes, which, in direct contravention of the rules of Lyell's own uniformitarianism, are not seen in action anywhere, now, or ever, to account for most of the work that we can see has been done. While we described and discussed some of these "unseen" agents and processes in the foregoing pages, there are many others that we ignored, including some that were invoked by the various writers we examined herein. We will, over the course of this and future volumes, deal with the others also, many of which do not even have so much as a theoretical basis for their claimed existence, let alone any evidence.

In virtually all cases, such unseen causes, or agents, formerly and disparagingly termed "transcendental" by uniformitarians themselves, are invoked to overcome the obvious inability of Lyell's uniformitarianism to account for the evidence. This incompetence was pointed out by many of the older geologists when Lyell first published his theories in 1830. One of these was Henry de la Beche, the first head of the British Geological Survey, and a man who began his geological career 10 years before Lyell. He expressed his estimation of Lyell's thesis, especially the part claiming that rivers cut out valleys, with the cartoon of fig. 8-14.

While modern-day geology remains adamant that such "piddling streams" could, if given enough time, carve out valleys and wear away the rocks of a country to the condition we now find them, they claim that geologists in general, when presented with Lyell's supposedly brilliant insights in the 1830s, immediately recognized the error of their ways and abandoned their misguided adherence to their superstitious beliefs in Noah's Flood, or any kind of catastrophe. To a man, they mustered their toes in line with Lyell's new genius-derived, and self-established, "rules of the game," and never looked back, or so we're told—mostly by the British.

Bless the baby! What a walley he have a-made!!

Cause and Effect

Fig. 8-14: Henry de la Beche's cartoon (circa 1830) shows a child
peeing in a wide-open valley to create a misfit stream, a tiny stream in
a large valley. Few, or to be more accurate about it, approximately no
geologists in the early 1830's would accept that the "piddling stream"
shown here could cut out the valley it's flowing in. And today, many
still would not, and neither would I. Image: Clary & Wandersee, 2010.

History, as I may have said before, is written by the victors,
and uniformitarianism has ruled the roost now for over 150 years,
since the 1870s, and not, in actual fact, since the week after Lyell
published his book in 1830, as the "victors" would have us believe.
It took 40 years and the deaths of all the older geologists to enable
that to happen, and it certainly wasn't because Lyell had the weight
of evidence on his side. It was, as we'll see, rather a case of his making a lawyer's argument in front of a friendly judge, and in a court
that didn't recognize the laws of physics, preferring, it seems, Lyell's
newly established "rules."

While I have not hidden my objections (at the very least) to the claims of the uniformitarians, I have, at the same time, tried to objectively analyze the evidence from Greece and the Aegean. We have, as is one theme of this essay, turned a rather critical eye on the interpretations of that evidence by the orthodox geological community, assessed it with regard to Lyell's uniformitarianism, and compared the official claims of its origin against our familiarity with physics and the nature and magnitude of the forces that are very obviously required to produce it.

Plain common sense tells us that a big boulder needs a large force to move it, and a big roundy boulder lying in a valley can only have been shaped and moved there by a comparatively big flood; it obviously cannot be moved by a succession of small floods when no single one of those floods can be shown to have an effect on that boulder.

And how do we explain it when that boulder is found on top of a hill, as I, and everyone else who lives here in New England, can find them scattered all over the hills? Ice won't do it because ice can't round-off boulders, and it can't flow uphill. Plus, these boulders are usually found isolated, and ice doesn't do that when it melts; it dumps moraines in piles of rough debris at the foot of the ice front, just as we saw earlier.

Moreover, these big roundy boulders are the very same as those found scattered all over northwest Europe and into Russia. The effort to account for them, without the involvement of water, led to the adoption of the Ice Age theory in the first place.

As Sir James Hall put it as early as 1812, in a paper attempting an explanation for the large boulders found scattered all over the country, published in the *Transactions of the Royal Society of Edinburgh*:

"It is in vain that a vast duration is ascribed to the influence of an agent, unless it can be shown that its action has a tendency to produce the alleged result. If it has a tendency to produce a different result, that difference would be augmented in proportion to the duration of the action. Now diurnal operations are everywhere found in the act of corroding and altering the forms alluded to, but they are nowhere seen to produce them. This class of facts, on the other hand, all conspire in giving probability to the hypothesis of a diluvian wave, which affords an easy explanation of all the large features of this country." (James Hall, 1812)

A "diluvian wave," which is a massive wave sweeping over the country, cannot be confused with Noah's rather tame, rain-derived Flood, and James Hall here exemplifies the opinion of most geologists of his day. According to Hall, one must first show that one's favored agent can actually do the job that has been done, as Greenough said and as we saw in the epilogue to volume 1. That is a very basic critical scientific requirement of the science of Geology, logically enough. And, obviously, if that agent can't do the work done, then a different agent must be found that does have the required power and ability.

James Hall was an associate of both James Hutton, the so-called "founder of modern geology" and John Playfair, the man who rewrote Hutton's book, *A Theory of the Earth*, in which he espouses what was to become Lyellian uniformitarianism. In fact, Hall, who was the only geologist among them, accompanied them both on excursions around the local countryside, and was even present when they went to view Siccar Point, Scotland, probably the most famous locality in geology (at least according to the British).

It was here that Hutton saw an unconformity and decided that it represented long ages (an idea he actually got from the Swiss geologist de Saussure's, *Voyages Dans Les Alpes* [1779, 1786] from which he cribbed most of his book [Carozzi, 2000] as noted earlier) and it is a location that almost every British geology student is taken to see, and be duly awed at Hutton's supposed genius and insight, all the better to indoctrinate that innocent student with Lyell's uniformitarianism and instill a firm belief in the amazing power of piddling streams.

Hutton's book, *A Theory of the Earth,* which I mentioned and briefly discussed earlier, and Playfair's so-called elucidation of it in his *The Illustrations of the Huttonian Theory* were generally ignored by everyone, until Lyell came along and resurrected and embellished them both as his "great" *Principles of Geology.*

In order to negate Hall's influence and divert attention away from him, it has often been (disingenuously) claimed by orthodox geology that Hall agreed with Hutton and Playfair. He did, but *only* with respect to the igneous nature of granite, and that's it. Despite being friends with both of them, he could not accept that everyday processes, or "diurnal causes," as he termed them, could possibly account for the battered and worn-down condition of the landscape, or the deposits to be found everywhere. He published two papers, both in 1812, on the subject, titled "On the Revolutions of the Earth's Surface" (I and II), one of which contained the extract included above, and which are entirely opposed to Hutton's theory, and they are no secret.

Essentially, Hall argues that while everyday causes, mostly rivers, may wear and abrade the rock surfaces they pass over, so long as they're carrying sand and stones, etc., they do not have the power to move large boulders or carve out valleys. I have made clear my agreement with Hall's judgement many times herein, and there were many like him. Instead of piddling streams, Hall declares his agreement with the continental geologists that only large floods could do that kind of work (Hall, 1812).

And that is exactly what most of the evidence we saw in Greece seems to suggest, and *that*, as we also saw, is according to orthodox, uniformitarian geologists, many of them "continental."

Chapter References

Andriesse, J. P., 1988, Nature and management of tropical peat soils, FAO Soils Bulletin 59, Food and Agriculture Organization of the United Nations, Rome, 1988.

Athanassiou, A. et al., 2018, Pleistocene vertebrates from the Kyparíssia lignite mine, Megalopolis Basin, S. Greece: Testudines, Aves, Suiformes, *Quaternary International*, 497: 178-197.

Carozzi, M., 2000, H.-B. De Saussure: James Hutton's obsession, *Archives de science et compte rendu des séances de la Société* / édités par la Société de physique et d'histoire naturelle de Genève Vol. 53, No. 2, pp 77-158.

Clary, R. M. & Wandersee, J. H., 2010, Scientific Caricatures in the Earth Science Classroom: An Alternative Assessment for Meaningful Earth Science Learning, *Sci. and Educ.*, Vol. 19, pp 21-37.

Delogkos, E. et al., 2018, Structural Geology of the Lignite Mines in the Ptolemais Basin, NW Greece, *Proceedings of the 14th International Symposium of Continuous Surface Mining.*

Diamantopoulos, A., 2006, Plio-Quaternary Geometry and Kinematics of Ptolemais Basin (Northern Greece): Implications for the Intra-Plate Tectonics in Western Macedonia, *Geologia Croatica*, Vol. 59, Issue 1, pp 85-96, Zagreb.

Georgakopoulos, A., 2000, The Drama Lignite Deposit, Northern Greece: Insights from Traditional Coal Analyses, Rock-Eval Data, and Natural Radionuclides Concentrations, *Energy Sources*, Vol. 22, pp 497–513, © Taylor & Francis.

Hall, James, 1812, On the Revolutions of the Earth's Surface, I & II, *Trans. Roy. Soc. Ed.,* Vol. vii, 1812.

Iamandei, S. et al., 2012, Identification of a fossil trunk from Achlada Mine (Florina, Greece) newly installed at the entrance of the coal mine, *Romanian Journal of Earth Sciences*, Vol. 86, Issue 1, pp 41-49.

Isler, E. et al., 2016, Geochemistry of the Aegean Sea sediments: implications for surface- and bottom-water conditions during sapropel deposition since MIS 5, *Turkish J. Earth Sci.* Vol. 25, 103-125, p. 103.

———, 2016, Late Quaternary chronostratigraphy of the Aegean Sea sediments: special reference to the ages of the sapropels S1–S5, *Turk. J. Earth Sci.*, Vol. 25, pp 1–18.

Johnston, W., 2001, Ireland's Peat Bogs, from "Travel Through Ireland Story" website.

Karageorgiou, D. E. et al., 2010, Development of Lignite in Crete, Comparisons of Basins, Possibilities of Exploitation, *Bull. Geol. Soc. Greece*, Proceedings of the 12th Intl. Cong., Patras, May, 2010.

Karkanas, P., 2018, Sedimentology and micromorphology of the Lower Paleolithic lakeshore site Marathousa 1, Megalopolis basin, Greece, *Quaternary International*, Vol. 497, pp 123-136.

Kurbatov, I. M, 1968, The question of the genesis of peat and its humic acids, *Transactions of the 2nd International Peat*

Congress, Leningrad (Ed. R. A. Robertson) 1:133-137. HMSO, Edinburgh, (cited in Andriesse, 1988).

Mantzouka, D. et al., 2019, Two fossil conifer species from the Neogene of Alonissos Island (Iliodroma, Greece), *Geodiversitas*, Vol. 41 (3).

Mavridou, E. et al., 2003, Paleoenvironmental interpretation of the Amynteon-Ptolemaida lignite deposit in northern Greece based on its petrographic composition, *International Journal of Coal Geology*, Vol. 56, pp 253-268.

Metaxas, A. et al., 2010, CO_2 Content of Greek Lignite: The Case of Proastio Lignite Deposit in Ptolemais Basin, Northern Greece, *Bull. Geol. Soc. Greece,* 2010, Vol. XLIII, No. 5, Proceedings of the 12th International Conference, Patras, May 2010.

Michalakopoulos, T. et al., 2014, Discrete-Event Simulation of Continuous Mining Systems in Multi-Layered Lignite Deposits, Proceedings of the 12th International Symposium Continuous Surface Mining-Aachen 2014 (pp 225-239) *Edition: Lecture Notes in Production Engineering* 2015, Ed: C. Niemann-Delius, Springer International Publishing.

Oikonomopoulos, I. et al., 2008, Neogene Achlada lignite deposits in NW Greece, *Bulletin of Geosciences*, Vol. 83, Issue 3, pp 335-349.

Spratt, Capt. T. A. B., 1865, *Travels And Researches In Crete*, Vol. II, Appendix IV: On the Geology of Crete, London, John Van Voorst, Paternoster Row.

Thompson, N. C. et al., 2018, In search of Pleistocene remains at the Gates of Europe: Directed surface survey of the Megalopolis Basin (Greece), *Quaternary International*, 497: 22–32.

Triantaphyllou, M. V. et al., 2015, Holocene Climatic Optimum centennial-scale paleoceanography in the NE Aegean, *Geo-Marine Letters*, Vol. 36, pp 51–66.

Widera, M., 2015, Compaction of lignite: a review of methods and results, *Acta Geologica Polonica*, Vol. 65, No. 3, pp 367-378.

CHAPTER NINE:

FOSSILS OF GREECE

Plant Fossils of Greece:

The lignite basins, as we've seen, are rich in plant material, particularly trees, and they are in a good enough condition to identify many, if not most, of them. The lignite basin on the island of Alonissos is particularly famous for certain fossil trees and is of special interest because of a certain pine tree found there (Mantzouka et al., 2019).

The pine in question, *Pinuxylon alonissianum* to give it its Latin name, is closely related to similar species found elsewhere, especially in Central America, Mexico, Arizona, and the U.S. west coast. It is not related to similar pines from Asia, and hence it cannot be argued that this pine arrived in Greece by migrating from the east, or from over the Bering land bridge, as is so often suggested.

The latter suggestion is always made with little consideration given to the actual practicalities of such a circuitous route across Siberia. Firstly, the Bering "land-bridge" was far to the north, and well outside many species' natural habitats, the only places they could grow. The plants must somehow move north, traverse cold climates, climb up and over mountain ranges, cross swamps, bogs and rivers, and then move back south again, and do all this in a general environment that no warmth-loving species, plant or animal, could tolerate.

Another tree, a swamp cypress by the name of *Glyptostrobus Europaeus*, is now extinct everywhere except in China and southeast Asia. It was formerly widespread in Europe, eastern North America, and even in Greenland, going extinct at the end of the Pleistocene, which is to say, the end of the Ice Age. This is rather curious and significant: how could a swamp cypress live on Greenland during the Ice Age, when none can live there now?

The other plant and tree fossils found in any basin are reflective of pretty much all the others, and include oak, alder, beech, fir, hemlock, among others, and many shrubs and plants such as lianas, sassafras, ivy, etc. Such plants imply warm temperate, cool temperate, warm and dry, or arid environments, and mixed forests to open shrublands and grasslands.

It is as though plants from many types of climate, and environment, are brought together here in these lignite deposits, and much the same situation obtains in the other basins. However, this is accounted for by supposing changing climate over long time periods, even though all the plants are mixed up together in the same lignite layers, meaning they're all the same age. Also, there's no evidence of such a changeable climate, or its possibility, apart from this inference from the plants of the lignite basins, which is what we're trying to account for, so this is just more circular reasoning.

In a paper on the vegetation and climate of northern Greece about 6 my ago, before what is known as the Messinian Salinity Crisis, when the Mediterranean Sea supposedly dried up, Bouchal et al. (2020), used palaeobotanical data to reconstruct the climate of the time. Leaf fossils and pollen taken from the lignites of the Ptolemais-Servia Basin (which we looked at earlier) were shown to indicate that various types of forest were present in the area, including riparian and mesic.

The first is simply forest along the side of river, which is the same thing as a river flowing through a forest and, being common the world over, do not really indicate anything, but such a proposed river serves the purpose of a transporting mechanism to bring the leaves, twigs and branches, etc. into the basins.

Mesic forests are those well supplied with moisture, especially temperate-climate types. Other types of forest, and other environments, are also identified from the lignite material, much as stated in the previous paper from Mantzouka et al. (2019). At the same time, we are told that lakes and swamps came and went, and layer by layer these lignite deposits built up over long ages.

As I've said earlier, how any area can repeatedly turn from swamp to lake to dry land to swamp to lake, again and again, and apparently, at the same time, also change from one type of environment to another is entirely beyond me, and I have made it clear that I don't accept the notion for a minute. As to how all these different plant types, necessarily from so many different environments, could all wind up in the very same lignite layers, in the same basins, over a huge area is also very problematic to explain, at least by any normal, everyday process, especially when all these different environments imply such different climates.

The fact that all these different plants from different environments and climates are all found mixed together in each layer implies that they all arrived into that layer at the same time, implying that they're all the same age and hence do not reflect any change in climate or environment. Their presence obviously has a much different explanation and I have to conclude that a *transport* and *depositional* origin for the lignite is closest to the truth.

Regarding trees and other plants formerly common to Greece and the Americas, it is an obvious necessity that these two regions had to be connected by land to enable the plants to spread from one region to the other. They cannot be presumed to have all separately evolved in both places, as that goes hard against all evolutionary theory, and, indeed, regardless of what one thinks of evolution, the idea also tends to defy common sense.

Plate tectonics and the separation of the continents cannot help the species commonality, as that separation supposedly occurred about 60 my ago, before many of those plants even appeared, supposedly in the Tertiary Period, and the Greek ones went extinct about 5 my ago, in the Pliocene Epoch. At the same time, there are today still plenty of plants, and animals too, common to both sides of the Atlantic, as we'll see when we get to the Atlantic seaboard, or if we all just look out my back window in Massachusetts.

Thus, some kind of a connection, by land, must have existed up to relatively recent times. Various suggestions along the lines of floating logs, windborne seeds, bird migration, and animal transport have been proffered over the years to account for these commonalities, but I'll show later that these have little or (mostly) no scientific merit, especially the one involving birds.

Prior to the arrival of plate tectonics theory, such floral and faunal relationships were explained by overland connections known as "land bridges" which comprised either continuous land, or island chains, or both. Considering that we mentioned relations between a number of trees in Greece, and both central American and Greenland species, it is noteworthy that such land bridges were considered to have existed. One is theorized as going across the North Atlantic from Ireland/Scotland and Scandinavia to Iceland/Greenland and on to North America, and another southerly one across the Central Atlantic, stretching from North Africa to the Brazil/Caribbean region.

These would certainly account for the Greek species having western relatives. And, as we'll see when we examine the Atlantic, there is indeed a "ridge" of elevated sea floor running from northwestern Europe to Iceland and Greenland that comes within a few hundred meters of the ocean surface. In the south, near the equator, there are projections off of both West Africa and Brazil and, of course, the mid-Atlantic ridge in between, which, if all were elevated a certain amount, might constitute land bridges across the Atlantic. The plants and animals could then cross fairly easily, cooler climate species on the northern bridge, while the southerly bridge accommodated the warmer climate types, and, of course, animals and plants could travel in both directions.

However, it's worth mentioning here that animals and plants do not migrate in an east-west direction; they only expand their territories in these directions when, or as, their population increases, which itself varies with the total food supply available. As we're all well aware, many animals do migrate a limited amount (their "range") north-south every year following the ripening food supply. And just to cover the subject thoroughly, some animals on the African savannah do appear to migrate in something of a circle following the food. Therefore, while they do technically migrate in an east and west direction, it is only temporary and incidental. Every year they wind up back where they started.

Birds, of course, migrate from short to extreme distances between winter-summer habitats, but, like all animals, birds do not, under any circumstances, migrate east-west, so I wish our scientists would stop invoking bird-transport of seeds across the Atlantic, or east-west across any ocean or sea.

No bird crosses the Atlantic in an east-west direction. All I could find was one goose that summers in Canada and winters in

Scotland or northwest Ireland, which is more a north–south direction, and simply the shortest route for that goose to get out of Arctic Canada for the freezing winter.

The only other reasonable suggestion put forward was that by H. E. Forrest in 1933, and that was that these animals and plants may, instead, have originated on a former central Atlantic continent, now sunk beneath the waves, a topic we've leave for later.

The Animal Fossils of Greece

Having looked at the geology of Greece and seen evidence for lots of action by lots of water, in gross violation of standard uniformitarian doctrine, and the trouble orthodox geologists have had in trying to account for that evidence, we have good grounds for concluding that standard explanations may not be the correct ones.

The lignite basins seem to lead us to the same conclusion, while also giving us some information on the plants that once grew in (or near) Greece and hence the climate that formerly prevailed in the general region. For the most part, the climate seems to have been warm temperate to sub-tropical but may have been temperate to cool at times, or the temperate-type plants may indicate an upland, slightly cooler climate.

Whether all the lakes and swamps of the lignite basins existed is highly questionable, but at least there must have been plenty of rainfall to sustain all that tree and plant growth. The climate of the Ice Age would presumably have been cooler but in general few authors consulted devote much, if any, time to it.

The other main indicator of climate is the suite of animal life that inhabited an area, and the period covered by the foregoing is generally the last (supposed) 10 my or so, and thus the fauna of Greece of this period are the mammals and various other animals of the later part of Lyell's Tertiary Period, and including, again, Lyell's own Pleistocene Epoch. Animals, of course, are good climate indicators, like plants, because they are also restricted to certain north-south latitudinal ranges, though not quite so rigorously as plants.

Greece enjoys a certain fame for the amount and variety of the fossils contained in its many deposits, most of which are in the lignite basins and river valleys, including a number we have looked at for other reasons. The most famous, and one of the most famous sites in general, is at Pikermi, named after a nearby village; the fossils collectively being referred to as the "Pikermi Fauna."

The following brief description of the fossil site is provided by *thefreedictionary.com* and taken from *The Great Soviet Encyclopedia* (1979). Like all Soviet-era references, it comes with the following amusing caution: "! It might be outdated or ideologically biased." One wonders how a strictly-informative, albeit Russian, scientific encyclopedia entry could somehow be ideologically biased. Western uniformitarianism, of course has no ideological bias whatever. The "outdated" suggestion is a common enough charge, often serving as a vague critique or, here, as a doubt-sowing pre-emptive strike.

As a matter of interest, we will meet a lot more "Soviet opinion" when we explore the Atlantic, especially since Russia, Soviet or otherwise, has explored and studied it far more than anyone else. To continue with Soviet opinion:

> "The Pikermi Fauna (name derived from the village of Pikermi, near Athens, Greece) is a group of fossil mammals that were widely distributed in southern and temperate latitudes (up to 55° N lat.) in the upper Miocene and the Pliocene in Europe, Asia and northern Africa. The first finds were made near Pikermi.

The origin of the Pikermi fauna was associated with the development in Eurasia during the early Neogene of grassy forest steppes, which were inhabited by various species of (three-toed) tridactylous horses (*Hipparion*), rhinoceroses, mastodons, giraffes, antelopes, deer, and other ungulates. Predators included civets, hyenas, martens, saber-toothed tigers, rodents, and apes. The fauna also included ostriches and other birds, turtles, lizards, and various amphibians.

The Pikermi fauna of various regions differed in genus and species composition, which may be explained by changes over time and by differences in physicogeographic conditions. In Europe and Asia, most representatives of the Pikermi fauna became extinct in the second half of the Pliocene, probably as a result of the cooling of the climate. In Africa and southern Asia, many descendants of the fauna constitute a significant portion of the extant mammals.

Large sites of Pikermi fauna have been found in India, Mongolia, and southern Europe. In the USSR, remains have been discovered in the southern Ukraine, Moldavia, the Caucasus, Middle Asia, Kazakhstan, and southern Siberia" (Trofimov, B. A., 1962).

The Pikermi fauna, therefore, has a major present-day African and Asian character, comparable to that seen in a savannah environment. Hence, on the face of it at least, during the Pliocene, this fauna used to live much further north than it does now, meaning Greece was much warmer than now. The fauna was also spread over a vast area of the Eurasian continent, and while, as the above encyclopedia entry states, there was some variation over this wide distribution, it was generally of much the same type, which is to say that of a warm savannah, presumably with rivers, lakes, trees, grass, and so on, much like many savannah regions of Africa and Asia, and including India, today.

The foregoing is a brief introduction to what is clearly a continent-wide fauna, named after the Greek locality because its remains were first found there in the mid-19th century. A more detailed description of the actual site, its exploration history, and its finds, is given by Roussiakis et al. (2019).

The fossil site is located in the Mesogea Basin, one of three major basins of Attica, and about 20 km east of Athens. A number of streams flow off of the surrounding hills and merge into the Megalo Rema river, one of the tributaries of which is the Valanaris. It is along this particular stream that, for some reason, most of the fossil sites have been found. The basin itself is just like all the others described herein, with conglomerates, gravels, sands, clays, marls, and, of course, lignites/coals.

The actual fossiliferous Pikermi Formation consists of a 30 m (100 ft.) thick variable sequence with silts predominating and subordinate conglomerates and sands. It lies "discordantly" (unconformably) on the layers below, implying erosion of the lower beds, above a limestone unit with palustrine (wetland/swamp) to lacustrine (lake) marls and coals, while it is "concordantly" (i.e., no erosion occurred) overlain by palustrine to lacustrine clay, coal and platy limestone. Therefore, it is sandwiched between two coal-bearing series which are almost identical, and the similarity with every other basin we examined is obvious.

The Pikermi Formation has been divided into two members, a lower conglomerate-rich layer which the authors tell us is: ". . . characterized by an alteration of red silts with a weak pedogenic overprint and debris-flow deposits." Thus, the lower sequence consists of massive silts, muddy material that the authors suggest is soil-like or soil-forming, and debris flows made up of conglomerates. This, again, sounds a lot like a repeating flood sequence.

It is within the lower Conglomeratic Member that most historic and recent excavations took place, and where most fossils have been found. The upper member, referred to as the Chomateri Member after the locality where it was studied, is, according to Roussiakis et al., composed of fluvio-alluvial silts with in-filled fluvial channels. The same fossils are found in the Chomateri Member as in the lower Conglomeratic Member, but the fauna is not as varied. Many new fossiliferous deposits have been found in the area in recent years (Roussiakis et al. 2019).

An enormous number of fossils have been found here, and Roussiakis et al. (2019) give a comprehensive list, in Latin, of the varieties listed by Trofimov (1962). Fossils have been mined out of these deposits on and off over the past 170 years, such that Pikermi fossils have by now been distributed to museums and universities all over the world, the fauna being represented as a "type-fauna" of the Pliocene. The fossil supply is, like the Siwalik Hills of India, and many other similar localities, nowhere near exhausted, which makes one wonder what killed them all, because this sounds like a mass-extinction event.

However, no one, to my knowledge, has ever mentioned a terrestrial mass extinction event at the end of the Pliocene epoch, only at the end of the Pleistocene, i.e., the end of the Ice Age, and even then, this latest one is not recognized as a bona fide "extinction event" to compare with the demise of the dinosaurs. A Pliocene *marine* extinction event is widely acknowledged, supposedly due to a supernova, though why the extinction was restricted to the oceans is not clear. In any case, we won't get into this question here.

Needless to say, animal fossilization involves the same necessities as the plant fossilization discussed earlier: the animal, or portion thereof, must be quickly and completely buried, deeply enough to be isolated from the open air (with its oxygen and bacteria) to ensure preservation. The perennial geological problem to be dealt with here is the same as usual, time and energy; while uniformitarianism says slow and gradual, and a little bit at a time over long ages, fossilization, to the contrary, requires rapid and deep burial by a lot of material, in short order, as exemplified by the polystrate trees we saw earlier.

Mammals are far and away the most numerous animal-remains found, and the list reads like some census of the African savannah at present, elephants, giraffes, rhinos, apes, horses or zebras, deer, antelope, wild pig, with a few extra types not to be seen today. What is notable, though, is that while the animals would all seem at home on the savannah, they don't all occupy the same "ecological niches" on that savannah, and one wonders how they all wound up together in mass graves in the Mesogea Basin of Attica.

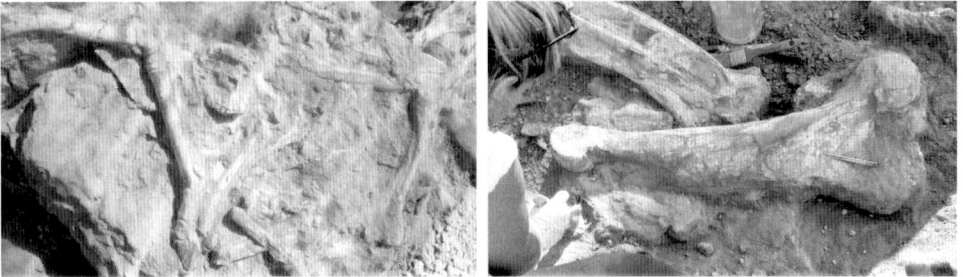

Fig. 9-1: On the left is a typical fossiliferous bone assemblage with in-situ articulated fore and hind limbs of three-toed hipparionin equids (pony or small horse) while on the right are limb bones of a proboscidean (elephant-type). The presence of multiple bones from the same animal, many articulated, as well as individual bones means that the animals were buried whole or in large pieces and all together in a layer. They must, therefore, all have been buried at the same time. Image: Roussiakis et al. 2019.

For example, most of the large animals live on the open range, while apes and pigs tend to favor forest or woody environments, as do some deer and elephants. In other words, here we have much the same issue as we had with the trees and plants from different environments in the lignites. Fig. 9-1 give an impression of the nature of the fossil deposits.

As noted, the fossils are found in silts with conglomerates, and the authors interpret the latter as resulting from debris flows. We have met debris flows, and their relatives, a number of times before in various other areas and contexts, often where conglomerates needed some sort of mundane explanation. In which case here, the vast numbers of animals entombed must have been entombed by said debris flows, or else they're not debris flows and instead are simply flood-deposited conglomerates, as we've seen elsewhere.

As we discussed earlier, it can be difficult to distinguish among tills, debris-flows, or flash-flood deposits, etc., but "debris-flows," which cover a very broad spectrum is, as far as explaining the fossils go, the least controversial, from the point of view of orthodox geology, that is.

Whether or not the conglomeratic member containing the fossils actually is, or even looks like, a debris-flow deposit is not the point. A debris-flow can easily be argued for because whatever buried the bones had to be of substantial thickness to bury them deeply enough for fossilization. However, the Pikermi fossils are found all over the Eurasian continent, and so the authors are hardly suggesting that debris-flows occurred over the same enormous area all at the same time, or at different times.

Roussiakis et al. (2019) do not tell us from where the "debris flows" came but they must have come from nearby mountains. On the other hand, Böhme et al. (2017) give a much more detailed description of the sediments, telling us that the red silts contain regular ". . . lenticular shaped sheets . . ." of conglomerates with a scoured base and are channelized at their deepest points. These 1.5 m (5 ft.) deep channels are bordered by what are usually 0.3 m (1 ft.) thick and 10 m (33 ft.) wide conglomeratic levees.

The clasts in the conglomerate, well-rounded and at up to 20 cm (8 inches) in diameter, definitely come from nearby mountains and the conglomerate sheets show an: ". . . indistinct inverse bedding (large cobbles on top) and their top (surfaces) can be highly irregular with an undulating surface and projecting cobbles." The inverse bedding is the reason for the debris-flow diagnosis as inverse bedding is usually ascribed to it, though debris flows are much the same as bottom slurries.

In some places in the formation, the conglomerate sheets were found lying one on top of another with some variation in texture being observable. The red silt, which makes up most of the member, contains some clay and fine sand, and also some floating pebbles up to 3 cm in diameter, especially in association with bone accumulations.

The authors interpret the silts as "entisols," which are simply masses made up of one material type, in this case silt. This means they show no sign of having anything to do with soil, and thus there's no sign of any former land surface, or swamp or bog, or anything suggesting plant growth; nor do these silts show any sedimentary structures like bedding, etc., or layering of any kind.

It is simply a mass of red silt with sheet-like lenses of well-rounded conglomerates and layers of bones. While the conglomerates in the silt must have derived from some nearby mountains, there is no indication reported that the silt did, and the fossils are found in the silt and, contrary to the Roussiakis et al. (2019) implication above relating to debris-flows, not in, under, or with the conglomerates.

Hence, the debris flows, if that's what they actually are, would seem to have little or nothing to do with the fossils, their deposition, or their burial. The red silt with the pebbles does, of course, remind us of Spratt's "red earthy gravels" on Crete.

Hence, it appears we may dismiss the role of "debris-flows" in the fossilization of the Pikermi fauna and wonder instead why there is no sign of the land, rivers, and lakes that the animals lived on and near. There is no sign, or any remanent evidence reported, of the presence of a savannah environment in the fossil area, just a mass of featureless red silt, with floating pebbles, and containing many dense layers of bones, with conglomerate lenses.

The standard explanation is that these animals were alive and well and living on the savannah, in the environs of lakes and rivers and swamps, into which they occasionally fell or got mired and drowned and were buried and eventually became fossils. In other words, they died and were fossilized (or simply preserved, as here) right near where they lived, and thus we should see some physical evidence of that environment. While not stated in the case of the Pikermi fossils, this is the usual uniformitarian explanation for any fossils found anywhere.

It is invariably an individual "accident," regardless of the variety and density of the bone deposits. Giraffe, rhino, bison or deer fossils, etc., are usually explained as an animal wading into a lake or swamp (for some unknown reason, as they can drink at the edge), getting stuck, and drowning. There were, apparently, no crocodiles in this particular well-watered savannah environment, which is decidedly odd.

One should either expect crocodile fossils in keeping with all the others, or else no fossils at all because the crocodiles would have eaten all the drowned animals. In fact, from what I've seen of nature documentaries on TV, it's usually the crocodile itself that's responsible for drowning the animals in the first place, immediately prior to making absolutely certain that that animal will never become a fossil, other than perhaps as a crocodile coprolite (fossilized turd).

So here yet again, present-day conditions and processes as we can observe on the African savannah are of no use in explaining the Pikermi fauna, and the conditions under which they are found, and hence we're missing yet another key. And, the utter lack of correspondence with present-day activities seems totally lost on any academic herein consulted.

However, while no crocodile bones were found at Pikermi, various types of cat fossils were, and these get the usual explanation. Although cats of most kinds typically do not go swimming, (tigers and jaguars being exceptions) saber-tooth, lion, or other fossils are explained as the animals drowning while trying to get at an already-mired prey animal; they must have somehow known that there were no crocodiles lurking almost invisibly, as they do, just under the surface of the water, or hidden in the reeds and bushes of these "swamps."

It's fairly safe to assume that the giant Pliocene crocodile (25 ft., 7.6 m long), being on its home turf and in its favorite element, would have come out on top of any altercation with any lion or saber-tooth foolish enough to have a go, and would hardly have balked at "coprolizing" either of them. (Jaguars, mind you, are noted for going after small caimans, but caimans are by no means in the same league as the giants of the Pliocene or even today's Nile crocodiles.)

Here at Pikermi, however, there is little sign of a normal savannah environment, and the fossils are all found in layers in an undifferentiated mass of red silt, that Böhme et al. (2017) explain as an entisol consisting of wind-blown dust/silt supposedly from the

Sahara, based on grain characteristics and salt content, the latter found in the sands of the Sahara. (We'll skip over, for the present, any question as to how salt got into the sand of the Sahara.)

However, the authors also conducted a microscopic analysis of the silt grains, and while the sand grains of the Sahara are exceedingly well-rounded, they tell us that the Pikermi silt grains are: ". . . mostly angular to sub-angular and have a high abundance of clays adhering to quartz grains, whereas some particles are composed solely of 'adhering clay aggregates,' which is typical of loess." The trouble here is that the European loess deposits are all to the north of Greece, while the Sahara is, as we know, well to the south. Plus, silt is widely distributed at all latitudes, not concentrated in the north or in the Sahara—see, for example, West Asia.

Fig. 9-2: Sharan desert sand, showing its pinkish color from iron oxide staining. The sand is composed of pure quartz grains which, as can be seen, are extremely well-rounded from their being blown and rolled along and impacting one another. Image: © Can Stock Photo/PixelsAway, Royalty-free Lic.

At the same time, the origins of both loess and silt are still subject to debate, and have been since the mid-19th century. Loess has long been considered to have derived by aeolian transport from Ice Age glacial deposits in the far north, though considering they were frozen or wet for most of the time, it's difficult to understand how so much material blew so far away from the polar regions, especially in the absence of adequate winds, which do not blow far from the arctic. In fact, Arctic and Antarctic winds blow in an easterly direction and are known as the "Polar Easterlies" and do not get very far out of the polar regions.

The issue with silt is that the typical particle size requires that quartz grains, newly released from a rock, would have to suffer a size reduction of over 90% to become silt, (Smith et al., 2002) and that is a tall order, because quartz is harder than most other minerals. Thus, most silt is considered to have formed by grinding in glaciers, based on the supposition that any glacier is a kind of super-efficient grinding mill. Since silt is also found much further south, in arid areas in and near deserts, and not to mention everywhere in-between, finding a process that will produce it there is a bit more difficult, even though the polar idea is baseless.

Smith et al. suggest a number of weathering processes, as well as both aeolian and fluvial abrasion and attrition during transport (see fig. 9-3). However, as just reported, the silt at Pikermi comprises angular to sub-angular grains, as well as a proportion consisting of "adhering clay aggregates," clearly indicating no abrasion of the grains, as should be expected during such long-distance transport, either from the Sahara or from polar regions. These adhering clay aggregates also indicate that that silt did not derive from the grinding of quartz grains in glaciers, or anywhere else. Clearly therefore, silt in general also has another, non-attritional, origin.

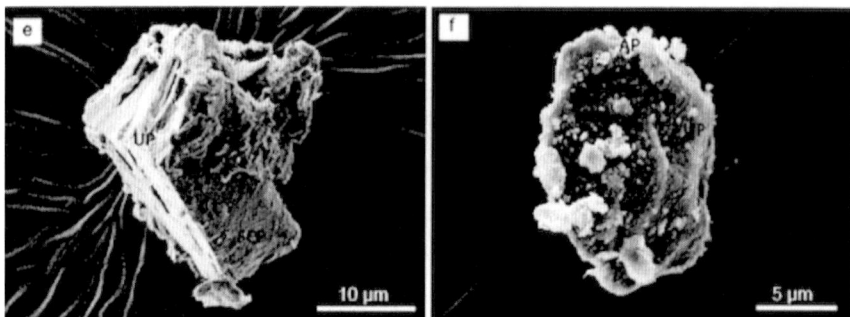

Fig. 9-3: Electron Microscope images of the silt grains of the Conglomeratic Member of the Pikermi Formation, showing angularity on the left and rounded on the right, while both have clay particles attached. The attached particles obviously imply no abrasion during transport. Image: Böhme et al., 2017.

We will be looking at these issues in more detail later, as they crop up in various contexts, both with regard to Atlantis and geology in general. Suffice to say for now that although the origin or mode of formation of both materials is still open to question, the fact remains that silt particles *are* found mixed with desert sand, and desert winds do spread silts similar(ish) to the northern loess over adjoining areas, though Saharan dust grains are cleaner than those of fig. 9-3.

Such desert-silt or loess is termed "peridesertic loess" (Smith et al., 2002). Now, as to how the animals found themselves in a mass of this "loess," the authors do not venture to guess, which is somewhat surprising, since a massive "dust storm" might seem to offer up a ready-made explanation for their demise.

However, quite apart from accounting for the death and burial of the animals, how is the massive bed of silt itself to be accounted for? If it blew up from the Sahara Desert as a dust storm, which is the implication, and it consists of one massive bed, and there's only one, then it means that at some point in the past, there occurred the most enormous dust storm the world has ever seen, and one that went on for ages.

In fact, Böhme et al. (2017) date the Red Conglomeratic Member to an interval spanning 7.37 to 7.17 million years ago, a time period of 200,000 years, which is one very long dust storm, with no interruption, as should be revealed by layering in the silt. At the same time, supposedly, the occasional "debris-flow" swept down off the hills, and all the while, again supposedly, vast herds of savannah animals were living happily in the neighborhood, eating and drinking and minding their own business, just like all those American farmers were doing on the Great Plains during the Dust Bowl of the 1930s.

While silt does blow up from Africa, I very much doubt that the sand or the 3 cm "floating" pebbles found in the silt came with it; in which case one wonders how these pebbles, especially, got mixed up in the silt and in such close association with the bone "accumulations," which also need to be explained. It would seem that whatever about the *origin* of all that red silt, its deposition could not have been one continuous process, but must have been interrupted on occasion, and it is obvious that water action was also involved, as evidenced by the debris-flows, bone layers, floating pebbles and sand/clay fractions, despite the lack of layering or any sedimentary structures, as Böhme et al. report.

They do tell us, however that a very similar deposit of red silt, also 30 m (100 ft.) thick is found at Curcuron in southern France, which is also rich in essentially the same fossil fauna as found at Pikermi, 2,000 km away, and is thus considered contemporaneous. This silt also supposedly came from Africa, though Africa is simply, and obviously, the only reasonable-seeming option available.

Quite apart from this evidence from France not contributing in the slightest to an explanation of the red silt, the presence of the same fossil fauna 2,000 km away would tend to suggest that similar "events" occurred all that distance away to produce those fossils. It was not likely the intent of these writers to provide supporting evidence of a regionally acting destructive agent, but that is essentially what is implied here, at least according to my interpretation, as repeatedly demonstrated herein.

In contrast to the Conglomeratic Member, the overlying Chomateri Member is a 7 m (22 ft.) reddish to yellowish series of silts, alluvial fine to medium sands and fluvial channels filled by medium to coarse sand with a minor content of pebbles up to 4 cm (1.5 in.) in diameter. The authors (Böhme et al. 2017) tell us that the: ". . . individual fluvial bodies (infilled channels) are interlocked to continuous laterally extensive channel-fill trains." This arrangement sounds like a braided stream network such as we saw on the surface of alluvial fans earlier.

Yet again we see that: "Three successive fluvial episodes can be distinguished . . ." while here at least, the authors report the presence of well-developed paleosols adjacent to the fluvial channels, so we finally have some evidence of a land surface. These authors then issue an interesting caution in that: "It is important to note that the lithologic characters of the Pikermi Formation, especially of the lower Red Conglomeratic Member, are very similar to the local Pleistocene cover and difficult to distinguish in the field, which has caused some misinterpretations in the past."

The differences are that the Pleistocene debris-flows are more dominant, by volume, over the fine sediments and contain clasts up to 1 m (3 ft.) in diameter, while the fine sediments do not contain vertebrate fossils. One can conclude, therefore, that the later Pleistocene sediments had a similar origin to the Pliocene Pikermi Formation. "Clasts" up to one meter in diameter are what we usually call "boulders," meaning the usual major flood operated here at one time.

In between the Pikermi (Pliocene) and the Pleistocene, is the Raffina Formation which is a typical lignite deposit. It consists of clays intercalated with lignites and coal as well as organic-rich clays. The sediments have been thoroughly mined for lignite, or coal, but immediately above the Chomateri Formation is a 2 m (6 ft.) thick layer of xylitic lignite beds and tree stumps in the upright position, and presumed to be in their place of growth. The beds also contain freshwater fossils such as *Planorbis*, *Theodoxus* and *Melanopsis*, all of which we met before, and which imply a lacustrine environment.

However, further to the east, the Raffina Formation is also exposed in a cliff face and displays a somewhat different composition compared to that at Pikermi, containing large numbers of marine fossils, primarily giant barnacles as well as some scallops and oysters (Dermitzakis et al., 2009). These authors here describe the Raffina Formation as consisting of fluvial conglomerates which pass laterally into (or are overlain) by a marine succession, mainly composed of conglomerates along with silts and sands. The giant barnacles (balanids) and other fossils were all transported and deposited postmortem, the transporting agent having swept these shells from offshore in over the land.

The authors suggest storm surges during temporary high stands of sea level (transgressions) which alternated with regressions, i.e., low stands of sea level, and repeated three times as "cyclothems" (repeating sediment series). The cyclothems are: "successively thinning upwards to suggest a waning influx of the returning overflood during regression."

Now, while such risings and lowerings of sea level are, as usual, supposed to occur imperceptibly slowly, "returning overflood" definitely connotes a much more rapid process, as is required to move large clasts and conglomerates, as discussed herein almost *ad infinitum* (or maybe *ad nauseum* at this point). This, again, sounds like a giant flooding and ebbing tide, which would mean that a flood, or floods, swept through the Mesogea Basin in an easterly direction, and then ebbed and flowed as a flood and returning overflood for a total of three cycles. And this sequence of events is more or less exactly what we saw all over Greece and the Aegean.

In explaining the whole Pikermi/Raffina series, Dermitzakis et al. propose that the low-land area, i.e., the Mesogea Basin, was a "polje," i.e., a basin formed by dissolution of limestone, while every other paper consulted considers it a tectonic basin, like all the rest, and with which I agree and reject the polje idea.

The eastern part of the basin lies south of Pentelikon Massif, and the authors state that the basin was filled during the Pliocene such that: "all karstic residue/debris have been 'scoured' from this massif by 'torrential' washouts to fill the marginal polje up to an even level by the decline of Pikermian sedimentation . . . explains both the torrential-fluvial and the temporary limnic conditions prevailing during the development of the whole Pikermi sequence, as observable at its classical localities, Megalo Rema and Chomateri" (Dermitzakis et al., 2009).

Presumably, the land was then flooded by a returning overflood in the form of a marine invasion which brought in the barnacles, etc., and this obvious east-west flooding was repeated three times, so what we're talking about here is simply a sequence of three ebb-and-flows of a massive flood. It seems that everything we see in Greece tells much the same story.

Fig. 9-4: Pentelikon Massif showing massive rough channeling on its particularly, and unusually, rough and uneven southern slopes. Image: CC-By-4.0 Int. Lic., photo Dimorsitanos (cropped).

Howsoever the basin came into existence is not very important, but Dermitzakis et al. (2009) do agree that the material filling the basin, i.e., both the Pikermi and Raffina Formations, to a great extent, was swept down from the Pentelikon Massif by torrential floods. They do not, however, offer a source of water for said floods, but presumably they would consider that it was Monsoon rainstorms as are normal in savannah regions, and are, no doubt, assuming we can make our own "leaps of faith" here.

However, this flood then swept on to the east and into the Aegean, and then, as reported, swept back as a marine invasion to bring in the barnacles. "Ordinary" torrential flooding originating on the Pentelikon Massif could not possibly attain the velocity to *sweep* out into the Aegean—and what, exactly, would bring the water back as a returning overflood, and with the velocity to sweep up and transport masses of barnacles and other shells?

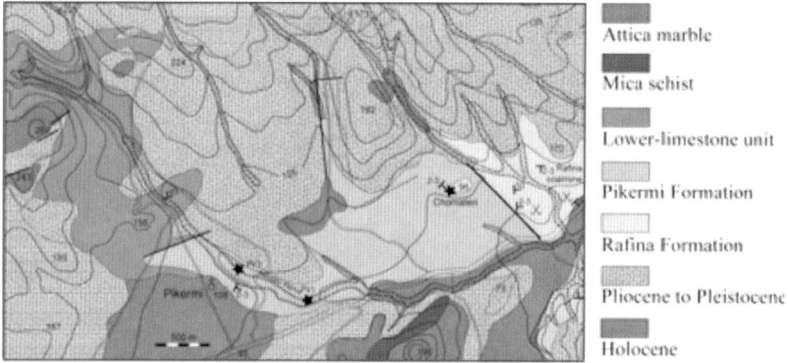

Legend:
- Attica marble
- Mica schist
- Lower-limestone unit
- Pikermi Formation
- Rafina Formation
- Pliocene to Pleistocene
- Holocene

Fig. 9-5: Geological Map of the eastern Mesogea Basin showing extent of Pikermi and Raffina Formations, and the river and stream courses (in notched lines) considered the routes by which the "debris-flows" and other sediments were swept off the Massif by supposedly "torrential floods." Image: Böhme et al. (2017).

The map of fig. 9-5 shows the locations and relationships of the Pikermi and Raffina Formations, while the grey contoured area is the Pentelikon Massif (also called Penteli Mt.).

As can be seen in fig. 9-5, the Pikermi and Raffina Formations are separated by a straight line, obviously a faulted contact, and with a small outlier of the Raffina Formation at Chomateri. The Pikermi is shown extending in a narrow tongue up a valley of the Massif, while also extending in a narrow belt around the base to north of Pikermi village.

It is more than likely that both the Pikermi and Raffina Formations are more extensive than shown here, and much is probably buried under later Pleistocene and Holocene sediments, shown in grey-shaded and plain dark grey at the bottom of the map. Thus, the filling of the eastern part of the Mesogea Basin, like many others, here again necessitates the invoking of torrential floods coming from the northwest.

With regard to evidence of the former environment, the authors (Böhme at al., 2017) offer only microscopic evidence taken from the silt itself. This evidence consists of phytoliths, which are the minute remains of silica minerals formed within plants, or they may be fossilized particles of plants, and are much the same as the fragments found in the cores taken off the coast of west Africa, referenced by Berger (2007) in volume 1.

The authors also use palynology, which is the study of pollen grains, and finally, micro-charcoal, which is self-explanatory. Phytoliths are commonly produced by grasses, and thus their finding is no surprise from a savannah, and both short and long grass types were identified, as well as herbs, shrubs, flowering plants, woody plants, and palm trees.

Quite a large proportion of pine trees as well as some oak were found in the silts. According to the authors, the phytoliths strongly imply an open grassland habitat with woody cover estimates of about 40% which is typical of a woody savannah, though the faunal remains found all over the region don't leave much room for an alternative. Both the animals and plant-remains found in the red silts imply a savannah environment, but it would be nice to have some evidence of the swamps, rivers, and lakes the animals supposedly died in.

None of the papers, books, or websites I've consulted offer any explanation for why the fossils are found all massed together in layers up to 30 cm (1 ft.) thick (Böhme et al., 2017, supplementary material) which they refer to as bone *accumulations*, as though the bones were gathered together over time, a process implicit in the meaning of the word "accumulation."

The same source tells us that the fossils occur in at least eight horizons and consist of "densely-packed bone beds," in which case one must ask why the animals decided to home in on this one particular spot to do away with themselves and pack their bones densely together. And if for each layer this accumulating went on for ages, why didn't the first bones rot, or get eaten, or get scattered by water or floods before more bones were "accumulated" on top of them? Not only that, why did the animals take seven breaks from this habit only to return after an interval to repeat the process?

Eight or more layers means eight or more double time intervals, one time-interval for each fossil layer and another for each intervening sedimentary layer, according to standard practice, and plain logic to a certain extent. The very essence of uniformitarian geology is that layers mean time, and the thickness of a layer represents, to a degree, the length of that time interval.

The fact that the silt, which contains the layers of fossil bones, shows no layering whatever itself, should give some pause, since it is at odds with the layering of the fossils. The origin of the silt as Saharan "loess," even if it has salt in it, is highly questionable given the presence of "floating" pebbles up to 3 cm (1 in.) and so it must have a different source, not involving wind.

There is, of course, no way these pebbles were blown across the Mediterranean from North Africa along with the "Saharan dust" and this explanation for the silt is, I'd venture to guess, based solely on the fact that dust blows from the Sahara at the present day, and thus it is a believable and uniformitarian explanation, despite the pebbly evidence. The lack of layering, and the minor content variation of the red silt normally and logically means a continuous but varying operation, while the layers of bones, by the same token, mean a discontinuous and intermittent operation, which is clearly contradictory.

Mentioning a similar silt and fossiliferous deposit from the south of France, as if it's also from Africa, seems to be merely some sort of implied "supporting evidence," when it's actually just evidence of the same inexplicable sequence of events happening 2,000 km away in France.

Oddly enough, in another paper (Theodorou et al., 2010) on which Roussiakis collaborated, the authors comment on this conundrum but first tell us that the Pikermi fauna is: ". . . commonly considered as chronologically and taxonomically homogenous, though the early authors already noticed that the fossils occur in at least two stratigraphic levels."

One early writer, Woodward (1901) noted two and, locally, three layers separated by 2 m and 1 m. The authors (Theodorou et al., 2010) suggest that older fossil collections may represent a fairly homogenous fauna, but only if the sedimentation and fossil accumulation had been very rapid, each one lasting a few thousand years and providing "snapshots of the fauna in slightly different time slices."

At which point one must ask why there is no layering in the silt to represent those "snapshots" and how would fossil accumulation have been rapid at times and simply non-existent at other times, when everything is supposed to have gone on at the same petty pace from day to day, just as Lyell said, and all his disciples preach. And we know animals die on a continuous basis, but do so usually at the hands, or rather jaws, of a predator, which then typically eats the animal, leaving little behind eventually, especially since the follow-up scavengers would take care of the leftovers, and hyenas and vultures, in particular, eat bones and all. Even the resulting coprolites typically don't get fossilized, as various insects and bacteria take self-nourishing care of them.

Orthodox interpretations of such fossil accumulations usually describe these phenomena as being the result of animals drowning in rivers or, as mentioned, getting mired, etc., and the bones are then washed down to some depositional site where they "accumulate rapidly." This, of course, requires that the bones suffer no deterioration while lying exposed by the side of a river, where anything can get at them.

This is another old "bone of contention" in geology; bones are resilient, but they are no different from any other organic materials. If exposed to rainwater, oxygen, animals or bacteria, they will naturally disintegrate or get eaten and disappear. It has been observed by many that of all the millions of bison (buffalo) killed on the great plains of North America during the 19th century, not a single bone, horn, or speck of a skull can be found today, a mere 150 years later, while another observation is that of all the thousands of stray dogs of New York city that were thrown into the Hudson River in that same 19th century, not a single one was ever brought back up in a dredge.

It is the same in any swamp, lake, plain, or savannah; if an animal dies, it disappears completely, and leaves little trace of itself, except maybe the odd horn or bone fragment, and the nourishment of the next generation of any life form that got a few bites out of it. According to what we know of current animal behavior, and the biology and chemistry of decay, it would seem that a completely different sequence of events occurred here at Pikermi. And we must wonder again at exactly how all those Pliocene plant and mollusc remains survived unmolested for so long in the sands and gravels of Alaska.

The fact that the bones are found in dense layers, packed tightly together, and with full and partial skeletons still articulated, suggests much more strongly that all these animals died together and were collected together in masses. They were then deposited in those masses where we now find them. It is not in the least bit surprising that no writer sees fit to compare these Pikermi "accumulations" with any of the similar and different bone deposits found all over

Europe and the Mediterranean region, and in much of the rest of the world also.

From Europe to Asia, these bone "accumulations" are found in caves, filling ravines, gorges, and rock clefts, even covering hilltops and in an untold number of basins similar to those we've seen in Greece, and many of them are massive, containing bones by the tens and hundreds of thousands or millions. And these bone deposits are precisely what William Buckland was writing about in his *Reliquiæ Diluvianæ* (*Relics of the Flood*) as far back as 1824, from which I extracted in the epilogue to volume 1.

Thus far, we have examined the older Pliocene fossils, typically dated to anywhere from 5 to 8 my old, based originally on Lyell's dating game, and all based on an original determination that the Pikermi fauna all represented a Pliocene one, and, therefore, was at least 2.5 my old, the actual age of the Pikermi fossils being (supposedly) up to 7 or 8 my, as we've seen (Bohme et al., 2017), though anything over 5.3 my is technically Miocene.

Above the Pliocene lies the Pleistocene, and, as we saw at Pikermi, it is directly above, and of a very suspiciously similar composition to the immediately underlying Pliocene deposits, to the point that they have been confused, as we were cautioned (Böhme et al., 2017).

The Pleistocene Epoch, according to orthodoxy, began about 2.5 my ago and ended with the end of the Ice Age at (precisely) 11,711 years ago (Niels Bohr Inst., 2008). Therefore, at Pikermi at least, the uppermost Pliocene formation, which is the Raffina Formation with its lignite layers, supposedly represents approximately 5 my of deposition until the Pleistocene was deposited on top.

Pleistocene fossils, like their Pliocene ancestors, have been found all over Greece, the Aegean, and neighboring regions. And the

Pleistocene fauna, as we'll see, is very similar to the earlier Pliocene one, even though, as we saw, that fossil Pliocene fauna indicated a savannah environment and the Pleistocene is supposed to have been a period of much colder climate with an Ice Age going on.

This is something of a puzzle, though most writers on the subject generally suggest various interglacial warmer periods (escape hatches) within the general Ice Age to account for what is otherwise highly anomalous evidence, though interglacials are just as much a mystery as Ice Ages are.

However, and playing along with orthodox theory for a minute, it is a further puzzle as to where these animals sought refuge during the colder periods of the Ice Age, since the only way to escape the cold is by moving south, and the Mediterranean is obviously in the way. We will address this conundrum when we have examined the fossil deposits themselves.

In general, Pleistocene fossils are almost all found in the same situations as the Pliocene ones, i.e., in the basins, as well as on islands in the Aegean and in a number of caves, the latter a very common phenomenon, as mentioned. Also, as well as being dispersed singly, the fossils are found in concentrated layers or other "accumulations," just like the Pliocene, and many basins contain multiple stratigraphic layers containing fossils in various concentrations. Again, and as before, whole and partial skeletons, as well as individual bones, horns, tusks and teeth have been found. Since we've already had a thorough look at the contents of various lignite basins earlier, we will here conduct a more limited examination of the evidence.

We've already met Pleistocene formations, which include lignite layers, in the Megalopolis Basin, in the Peloponnese, and, as we recall, Karkanas (2018) concluded that the lignites and other sedimentary materials originated as flood-, or "hyper-concentrated flow" (liquid mud/silt/sand) deposited layers, quite unlike everybody else's explanation for the lignites in all the other basins. Hence, going by Karkanas' diagnosis, everything found in this basin, in the actual lignite seams or elsewhere must have been deposited by those same floods or flows.

At the same time, others, e.g., Athanassiou et al. (2018) whose work we will examine shortly, consider the sedimentary infill of the Megalopolis to be ordinary lacustrine sediments. However, this author is referring to the northwest end of the basin, at the Kyparíssia mine, which, as shown in fig. 8-13, is almost entirely lignite.

In the central area of the basin, the presence of fining-upward series of gravels, sands, and silts, and others of just sand and silt, as reported by Tourloukis et al. (2018) would tend to negate that claim, and not to mention the floods and hyper-concentrated flows of Karkanas (2018). Further, gravel, sand, and even silt are not typical lake (lacustrine) sediments, since there is no mechanism acting anywhere in or into any lake that can distribute such sediments in a body of still water.

Any river flowing into a lake dumps all its heavier sediment at the edge, just like a river meeting the sea, as we discussed earlier. The only sediments one finds in a lake are muds or marls, and the latter only if the feeder rivers flow through a limestone area, to dissolve and bring in calcium carbonate. It is clear that the lake is drafted in here to supply the necessary environment and water-source for the animals to live in and around, and of course, to drown in and be deposited in, and somehow (never stated) to be rapidly, and necessarily, deeply buried in, before they rot away to the usual absolutely nothing.

The Megalopolis basin has long been known for its many Pleistocene fossils, but it is only recently that excavations have been undertaken. A number of workers have explored the basin from a paleontological and ecological view. Athanassiou (2018), for example, reports that the fossil finds in the Kyparíssia mine derive mainly from organic-rich sediments (lignites) and two fossiliferous layers (horizons) are present.

The animals found include hippopotamus, elephant, a number of deer species, beaver, three-toed animals such as zebra, tapir and rhinoceros, horse, hyena, lion, panther and ruminants such as bison or auroch. Despite the presence of elephants, hippos and lions, etc., he suggests a temperate woodland-forest environment with the (by now) usual body of open water. The animals, on the other hand, seem to indicate a savannah environment with an admixture from more northern regions.

A second paper on the topic by Athanassiou et al. (2018) covered various other types of animal including pigs or wild boar, birds, such as swans, darts and other aquatic types, and turtles. In this paper the environment is again claimed to be lacustrine, but it is also claimed that it was alternating with a fluvial one up to the middle of the Pleistocene. Hence, the infill of Karkanas et al. can be accounted for, supposedly.

However, we looked at the Megalopolis Basin earlier, and the section in fig. 8-13 shows a very complicated arrangement of lignite layers with multiple and uneven seams, dispersed and partly discontinuous over the basin, and all generally thickening and merging together toward the northwestern end, and lying on and against the bedrock below, with the Kyparíssia lignite almost one solid mass.

Considering the almost solid and considerable thickness of the lignite in the north-west, orthodoxy would imply that there has been no change in conditions in this area throughout the entire duration of the supposedly long period of the Pleistocene that the lignite layers represent, and, while I can find no complete timeline for the full series in the basin, Tourloukis et al. (2018) give us a period of 500-600 thousand years (dating it to the period 0.9 ma to 0.35 ma ago) for the infill lying below Seam II, while presumably the rest (Seam II to the top of Seam III) took the last 300-400 thousand years to accomplish.

This is a time period of 800 thousand to 1 million years, but there is no sign of the thousands of layers of other sediments that should represent that time period, just as we've not seen elsewhere.

With the fossils all found "accumulated" in two horizons, and with many whole and partial skeletons, and at a high fossil density, or concentration, it would seem much more likely that these animals were all deposited by the same floods that laid down the lignite seams, while ignoring, for the present, the mechanism or agent by which they met their demise. In any other context, the science of geology considers everything in a layer, or stratum, to be stratigraphically of the same age; it is more or less a bona fide "principle" and, indeed, one that follows from logic.

The question then simply boils down to how long it took the organic sediments to be deposited and the lignites to form, and what, exactly, was it that rendered the wood of the lignites in the condition we find it? As we saw earlier, the nature of the Greek lignites is such that it seems some violence was involved, as they are all composed of crushed and ground-up wood, in the form of branches, wood chips, etc., and the preservation of whole trees, apparently still in their growth position implies they were buried suddenly and deeply, and that could only be done by large floods, which seems to be the case here in the Megalopolis also.

As for trees in the upright position, the fact is that most tree types found do not grow in swamps, and I did mention that a common feature of large floods is upright trees floating along, kept vertical by the weight of the root ball. The layer arrangement, the widespread distribution of the sediments and lignites, and the evidence of moving water, as implied by the gravels and the use of the word "fluviatile," all conspire to suggest that the infilling of the Megalopolis Basin was accomplished by multiple large floods that swept over the basin, and there never was any massive swamp occupying that basin at any time.

And these floods transported masses (mats?) of macerated plant material, branches, and whole trees floating vertically, and were very likely responsible for the deaths of the animals in the first place, and all these dead animals were bloated and floated either below the organic mat or on top of it, or mixed in, or all of the foregoing.

Generally, the same situation obtains in various other basins where Pleistocene sediments and/or fossils are found. The situation, in fact, is very similar to that of the underlying Pliocene series we examined in many of those basins.

At the far northern end of the country is the Mygdonia Basin, in Macedonia, and many excavations have been carried out over the last few decades. This basin is so rich in fossils, which have been excavated since the 1970s, that it is, like Pikermi for the Pliocene, considered as one of Europe's reference regions for the study of Pleistocene mammals (Konidaris et al., 2015). It is also a lot easier and more comfortable to excavate here than in some dark, dank or stifling cave somewhere.

The basin fill consists of three formations, the lowest of which lies directly on the bedrock basement, implying that something eroded that basement clean of all sedimentary infill prior to deposition of this lowest formation, as we saw before. Three formations are again identified, as well as three fossiliferous horizons. The formations, like the others we looked at, consist of loose conglomerates, gravels, sands, silts, and clays, and are overlain by marls and marly limestones, with repetitions of the coarser sediments above the marls and alternating beds of gravels, coarse sands, silts and clays, which are described as "fluvio-terrestrial."

They are, of course, exactly the same types of flood, and repetitive flood, sediments we've already seen in almost every basin we looked at in Greece. No lignites are mentioned by any of the workers, and, therefore, the basin seems to have infilled by straightforward sedimentary deposition by moving water, but the marls, having to be explained as lacustrine, call for the usual changes in climate or tectonic movements or something else to explain their presence.

The lacustrine origin (directly from Lyell, in that he noted marl in lakes near his home estate of Kinnordy in Scotland, and true up to a point) is the only recognized uniformitarian and ongoing process observable today, as in: streams dissolve calcium carbonate as they flow through limestone districts and deposit it in lakes (or the sea) when the water becomes over-saturated. In many cases, these marls turn to rock very quickly, often only in a matter of hours or days, as with beach rock.

As a rock cement, calcium carbonate can harden very quickly, and marls are simply composed of pure calcium carbonate, typically with an admixture of mud, micro-fossils as are also found in the ocean, plant matter, or other impurities. What is notable is that no worker mentions micro-fossils in the marls of the lignite mines. That those marls might also be deposited by some other agent or the same agent that was responsible for the rest of the sediments, therefore, does not enter the mind of any worker, even though there is no limestone source for marl near many of the basins, nor in any of the river valleys leading to them.

Ultimately, that is the problem here; there is little to no limestone rock anywhere near the Mygdonia Basin, apart from some minor layers in rocks to the west and isolated outcrops of marble to the north of Thessaloniki, and marble is much harder to dissolve than limestone. The basin is surrounded mostly by gneisses, schists, and granitic rocks.

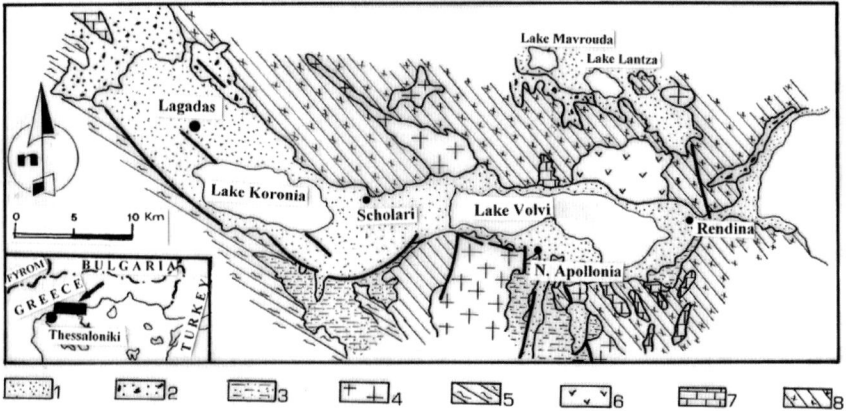

Fig. 9-6: Geological map of the Mygdonia basin. 1. Alluvial deposits, lacustrine sediments, deposits in river and torrent beds, valley deposits and a loose terrace system 2. Clays, sands and terrestrial phase conglomerates. 3. Marls, red clays and sands of lacustrine and terrestrial origin. 4. Acid to intermediate intrusive rocks. 5. Schists, quartzites, limestones and mafic rocks. 6. Amphibolites. 7. Limestones and marbles. 8. Gneisses and schists. Image: Miltiadis Nimfopoulos (2002).

Since, therefore, no rivers could have flowed through limestone to get to the Mygdonia Basin, the marl must clearly have a different origin, which it does. For now though, I will merely state that the source, or origin, of the marls in the Greek lignite basins is revealed, as one instance, when we discuss those "anomalous" plant fragments found in the cores taken by Hans Pettersson off West Africa in 1947.

According to Konidaris et al. (2021), the large mammal assemblage is rich and diversified, with saber-tooth, horse, mammoth, various bovids, deer, rhinos and giraffes, bears, wolf-like dogs, and giant hyenas. Birds have also been found, as well as reptiles. The fossils are found in the usual accumulations which include some articulated bones, and, as mentioned, in three different horizons. Most fossils are found singly, though, scattered through the sediment and spread widely over the basin. Fig. 9-7 shows a map of the basin with fossil localities.

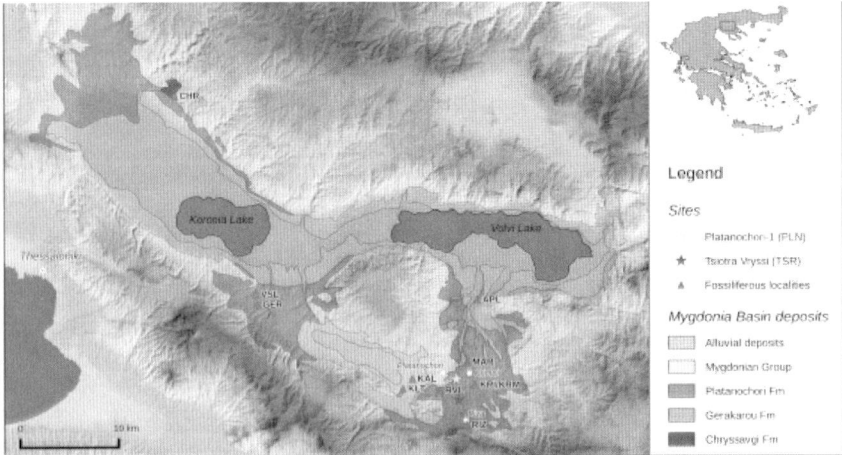

Fig 9-7: Mygdonia Basin to the west of Thessaloniki.
Image: Konidaris et al., 2015.

If any of the animals drowned in a lake and fell to the bottom, it is difficult to understand how their skeletons could have been scattered over the surface of the lake bed, while, as in any lake, the water does not move. And, as stated more than enough times already, it is an absolute necessity that the bones be buried rapidly and deeply to preserve them, as any *orthodox* description of the process of fossilization states, though you would never think it the way said orthodoxy so casually discusses fossils and fossilization.

Apart from the mere two basins examined, there are many more Pleistocene fossil localities in Greece and on its islands, containing mammoths and mastodonts, dwarf elephants, a record-breaking straight-tusked elephant, as well as many other types. It is noticeable that the (savannah) fauna of the Pleistocene differs little from that of the earlier Pliocene, despite the supposed change in climate to an Ice Age.

Greece, apparently, was teeming with elephants of one kind or another over both periods, as 75 elephant fossil sites have been discovered (Doukas & Athanassiou, 2003) some of which produced complete skeletons, indicating the animals were buried whole, before, during or shortly after death (Tsoukala et al., 2011). Leopards, lions, lynx, large wild cat, fox, badgers, rodents, wild goats, aurochs, hippos, lizards, extinct palaeoloxodon (ruminant) along with those already mentioned, and many more, have been found all over Greece and the islands, in basins, caves and various other situations.

The fossils are usually found in layers and at certain horizons, which indicates, according to standard stratigraphical principles, that the fossils in each layer are all of approximately the same age. This, on the face of it, means that the dying and burying of the animals was intermittent when that can't be the case under any circumstances. Either that or the dying was continuous, as it should have been, but the fossilization process was intermittent, which doesn't make much sense. The fossils are equivalent in age with those found in corresponding horizons, no matter how discontinuous or widespread, so clearly normal processes were not in operation in Greece at this time.

Since a layer of gravel or conglomerate cannot be deposited slowly, bit by bit, a pebble at a time, over an extended period of years or decades, the evidence suggests that major floods swept through these basins, and it was these floods that killed the animals and tore apart their corpses, just like they shredded the trees of the lignites, and deposited all those bones and skeletons in layers in the basins, each set of layers marking an ebbing flood (or tide).

An immediate returning flood then buried each fossil layer deeply under an overlying layer of sand, gravel, silt, etc., and rapidly repeated the process a number of times, thereby excluding oxygen and bacteria. That is a much simpler explanation for the evidence, despite conflicting with orthodoxy. It does, however, comply with the evidence and with logic.

This induction of mine would make explaining the fossils very much easier: the animals were all drowned by a flood, which tore apart many of them and then deposited them all together along with conglomerates, sands, and silts, etc., burying them suddenly and deeply in the absence of oxygen. However, this would imply a catastrophic end to these animals, a mass extinction, in essence, or at least a mass kill-off, and clearly more than one, but that is what the evidence implies.

At the same time, of course, it means that we would have no need of Lyell's guesstimated ages of millions of years to account for their supposed slow and steady demise. In which case, it is quite possible that the evidence we've seen in Greece bespeaks a number of major floods and catastrophes that may have occurred much more recently than any of the priests of *our* temples realize, or acknowledge.

A Note on Glacial Refugia

We have seen that the Pleistocene and Pliocene faunas of Greece were very similar and both more reminiscent of the savannah than of Greece today, and certainly as it must have been during the Ice Age. Large numbers of fossils have been found, most typically in the sedimentary basins, all of which show evidence of major flooding, despite what orthodox geology claims. It is the Pleistocene that is of most interest, because Greece, having had glaciers in the mountains during the Ice Age, as we're told, must have had a temperate to cool-temperate climate.

Such a climate implies cold winters with coolish springs and autumns, and savannah-type animals cannot live in such a climate, and certainly not as far north as 55°, despite what Soviet Trofimov said. It is not merely the cold but the lack of food during the winter, when there are no leaves and grass doesn't grow, as we all know.

Some larger animals can survive the temperate winters, e.g., deer, auroch and some other bovids, as well as tigers and other carnivores, for example. However, elephant, giraffe, rhino, hippopotamus and many other warmth-loving herbivorous and predatory animals cannot live outside the tropics, nor can many amphibians or reptiles, especially, and this presents another problem.

This problem is not confined to Greece by any means, but is met with in many areas of the world, particularly in the northern hemisphere, where we find lots of fossils of animals that are utterly unsuited even to the prevailing modern-day climate, and yet were supposed to have lived in the area during the much colder period of the last Ice Age.

The usual explanation has two components to it: one is that the Ice Age was not continuous but was broken up into a series of "glacial" and "inter-glacial" periods. During the inter-glacials when the climate ameliorates to a warmer phase, the animals come in and take up residence.

In the case of Greece, this must mean that the inter-glacial climate of Greece was much warmer than today, which does not make sense from climatological, meteorological, and astro-physical points of view. When the climate cooled again to a glacial one the animals migrated out of the region, and logically enough, to the south, where the climate was warmer and the environment, food supply, etc., would be more to their liking, i.e., they would follow the savannah southward.

Obviously, for the tropical animals of Greece and other areas of Europe, the big problem with this survival strategy is the barrier to said migration known as the Mediterranean, if, indeed, the Mediterranean Sea existed at the time, or at least in its present form.

Whether it did or it didn't doesn't matter, as uniformitarian geology, fortunately, provides suitable escape-hatches, called "refugia," for use in these awkward situations. Such physically impossible, and, therefore entirely imaginary, escape hatches are essentially "islands" of warmer climate existing in the midst of areas of colder or glacial climates. In other words, an "island" of a southern climate somehow (always unspecified) managed to exist in the midst of a northern one, and much further north than that climate could normally prevail, like some kind of Shangri-la hidden in the Himalayas, or something.

This, almost needless to say, goes entirely against the known astro-physical constraints on our climate system, entirely dependent as it is on the Sun-Earth geometrical relationship.

The idea is that typically southern species of plants and animals could survive the Ice Age in such "refugia," thereby "explaining" the presence of fossils of southern species found in northern deposits, and hence, we have yet more circular reasoning. Thus, the concept of "refugia" is based entirely on these out-of-place fossils and no other evidence. The concept itself is the opposite of the types of isolated climatic islands we do actually find at present.

Today, however, we only find them in warmer regions, such as the tropics or subtropics, where high mountains exist. The climate in the lowlands is warm, but as one climbs higher into the mountains, the local climate becomes more and more temperate with increasing elevation, due to the naturally cooler temperatures, and microclimates of different types exist at different elevations. Probably the best examples today are found in the high valleys of the Andes Mts.

It is, of course, perfectly logical and easy to understand how this situation can develop. It is extremely warm in the tropics and thus one is starting out with lots of heat at sea level and a hot sun shining directly overhead. Temperature declines with elevation at a more or less standard rate, due mostly to reducing air pressure, and depending on whether it's cloudy or clear, of 3-6° F per 1000 ft. or 6-10°C per 1,000 m. Therefore, in the tropics, where the ground temperature is high, there is a large quantity of heat present, and abundant rainfall at all elevations, any number of microclimates can develop at different elevations, all of which will be cooler than that at sea level.

As one moves away from the tropics the prevalence of cooler micro-climates at higher elevations declines with declining temperature, such that by the time one gets to Lat. 60, it's really only different degrees of cold, while at the arctic circle, there's only one frozen arctic climate at all elevations. Thus, we have no problem accepting the existence of what we may term "cool refugia" in the midst of warm regions, since all we have to do is park them up in high mountain valleys.

The converse, however, is not possible, because in a glacial or cold region one is starting out with a regional lack of heat, and without some way to convey a constant and concentrated supply of heat to a so-called refugium, that refugium cannot exist. Without heat, sunlight and water, the southern animals and plants that, by definition, give that refugium its name cannot possibly survive. And these animals are typical of the savannah regions of Africa and Asia today.

Further, it's not just a matter of heat; many plants have very specific sunlight requirements, such that they can only grow in very restricted latitudinally-defined areas. Glacial refugia, as they are envisaged for the Pleistocene fauna and flora, comprise three areas of limited extent in Spain, Italy, and the Carpathians, as shown by Sommer and Nadachowski (2006) in fig. 9-8.

The main problem with all of this, of course, is orthodox geology's insistence on ignoring the fact that all the Ice Age animals found as fossils, very obviously died by the millions at the end of the Ice Age, and not during it. We recall Baron Cuvier and his description of the animals found as frozen carcasses in the far north. He, and many since, have made it clear that the climate was completely different when the animals were alive, as it had to have been.

However, and very interestingly, the authors here tell us that: "The general problem with all genetic (origin of refugia) studies remains that the identification of refugia is dependent on hypothetical models, that are rarely referenced to data about the species contemporary geographical ranges" (Taberlet et al., 1998).

Fig. 9-8: Glacial refugia of mammals, southern Europe.
Three limited areas where large, warmth-loving animals
such as hippo, rhino, etc., survived the Ice Age to go extinct
afterward. Image: Sommer & Nadachowski (2006).

So, in other words, in their efforts to explain how warmth-loving species of animals and plants survived the freezing, food-less Ice Age, our scientists have constructed models that completely ignore present-day conditions regarding the various species' natural environments and their geographical ranges, which are exactly those critical data that tell one all one needs to know about an animal's favored and necessary temperature regime and food requirements.

Given that many of these fossil animals have very similar modern equivalents, it would seem the absolute pinnacle of logic to study those present-day animals first, to establish a baseline from which to work. This would naturally include their current behavioral patterns, migratory habits, reproductive needs, food and water requirements, and general survival strategies of those species that survived unchanged, or the modern analogues of those that didn't, if such exist. One must wonder what the scientists did, in fact, use to construct their "models," and why on earth they ignored the first rule of uniformitarianism, as well as abandoning logic before they even started.

In another paper on the refugia of southern Europe, G. N. Feliner (2011) discusses the very concept of refugia:

> "The concept of glacial refugia has traditionally been covered with a veil of mystery. It is suggestive of sanctuary, that is, a place where biological diversity has been preserved from extinction. Also, it is usually the case that refugia for specific groups are only imprecisely known, so that they have yet to be found or discovered. Despite this veil, they have a highly applied interest in conservation policies since an area that has facilitated long-term survival of species and populations throughout several climatic oscillations seems to be a good choice to be designated as a protected area in the long-term."... "Altogether searching for glacial

refugia is somehow like looking for the holy grail of evolution from the last three million years" (Tzedakis, 2004).

In the paper by Tzedakis referenced here, and concerning the Balkans as a refugium for temperate trees, this author makes a further point when he writes: "Indeed, when reviewing the large number of publications on this issue, one is often struck by a tendency (if not a desire) to infer the presence of refugial populations on the basis of sometimes indirect and even tenuous evidence" (Tzedakis, 2004). Here again we have an example of a questionable theory that must be maintained in the face of what seems a complete absence of evidence.

It remains, therefore, something of a major mystery as to how all these savannah animals, subtropical trees and the other plants we've met in Greece could have survived the extremely cold temperatures of the Ice Age, only to all go extinct at the very end. Indeed, according to some of the scientists involved, it seems easier to find the Holy Grail.

One aspect of the problem that has not been acknowledged by anyone, is that in the present day, few of those savannah animals of Ice Age Greece, and the Pliocene that predated it, could survive in the climate of Greece, and certainly not during the winter. They might be able to survive the summer, but Greece would need to be at the northern reaches of their migratory range, where the animals would, over the course of a year, migrate north and south following the ripening food supply, which is to say the animals' geographic ranges, which, as Taberlet et al. (1998) report, are the one thing completely ignored by the very scientists studying the question.

Before we ever explain their demise then, we first need to explain what all those savannah animals were doing in Greece and elsewhere in Europe to begin with. We will leave this question for later, but the situation in Greece and Europe is by no means as mysterious as what we'll find in many other places, phenomena that defy any effort at a uniformitarian explanation.

As for biologists actually studying animals in the wild, in a recent newspaper article on turkey vultures moving beyond their accustomed ranges and into the northern United States, the biologist writer admits to a general puzzlement as to why, and speculates on increased traffic and hence more roadkill (Pfeiffer, 2023).

The obvious answer I would think is global warming; animals and plants naturally move north, following their favored temperature regime. In addition, global warming, along with more CO_2 in the atmosphere, naturally leads to an increase in plant growth and, hence, a greater food supply further north.

Since it is now warmer in northern areas, the northern limit of the vulture's range has likely moved north accordingly, hence the vulture's presence. It is also quite likely that the vulture is no longer to be found at what was formerly the southern extreme of its range, due to its now likely and naturally (due to the same global warming) being too warm there.

Global warming and extra carbon dioxide in the atmosphere along with the fact that environmental conservation and restoration efforts in the last few decades have led to a population explosion among the wildlife of northern areas mean that, increased traffic or not, there is now more food available in a warmer environment. And this warmer environment is now such as the vulture was formerly accustomed to further south. Its range has, therefore, moved north due to global warming, quite logically.

Now, having examined many aspects of the geology of Greece, excessively at times, no doubt, and the theories that purport to explain it, and the typical attitude of our scientists, it's time to get back to the old Egyptian priest and Atlantis.

Chapter References

Athanassiou, A., 2018, Pleistocene vertebrates from the Kyparíssia lignite mine, Megalopolis Basin, S. Greece: Rodentia, Carnivora, Proboscidea, Perissodactyla, Ruminanta, *Quaternary International*, 497: 198-221.

——— et al., 2018, Pleistocene vertebrates from the Kyparíssia lignite mine, Megalopolis Basin, S. Greece: Testudines, Aves, Suiformes, *Quaternary International*, 497: 178—197.

Berger, W. H., 2007, On the discovery of the ice age: science and myth, in: *Myth and Geology*, Piccardi, L. & Masse, W. B. (eds.) *Geol. Soc. Lond. Special Publications*, 273, pp 271-278.

Böhme, M. et al., 2017, Messinian age and savannah environment of the possible *Graecopithecus* from Europe, *Plos One*, vol. 12, issue 5: e0177347.

Bouchal, J. M. et al., 2020, Messinian vegetation and climate of the intermontane Florina-Ptolemais-Servia Basin, NW Greece inferred from palaeobotanical data: how well do plant fossils reflect past environments? *R. Soc. Open Sci.*, Vol. 7, Iss. 5.

Dermitzakis, M. D. et al., 2009, Lower Pliocene (Zanclean) regressive sequence of Raffina near Pikermi in Attica, Greece: A spectacular locality of mass-aggregated giant balanid cirripedes, *Hellenic Journal of Geosciences*, Vol. 44, pp 9-19.

Doukas, C. S. & Athanassiou, A., 2003, Review of the Pliocene and Pleistocene Proboscidea (Mammalia) from Greece, in: Reumer, J.W.F., De Vos, J. & Mol, D. (eds.) - *Advances in Mammoth Research* (Proceedings of the Second International Mammoth Conference, Rotterdam, May 16-20 1999)-DEINSEA 9: 97-110 [ISSN 0923-9308].

Feliner, G. N., 2011, Southern European glacial refugia: A tale of tales, *Taxon* 60, April 2011: 365-372.

Forrest, H. E., 1933, *The Atlantean Continent*, sub-titled: *Its Bearing upon the Great Ice Age and The Distribution of Species*, H. F. & G. Witherby, London.

Karkanas, P., 2018, Sedimentology and micromorphology of the Lower Paleolithic lakeshore site Marathousa 1, Megalopolis basin, Greece, *Quaternary International*, Vol. 497, pp 123-136.

Konidaris, G. et al., 2015, Two new vertebrate localities from the Early Pleistocene of Mygdonia Basin (Macedonia, Greece): Preliminary results, *Comptes Rendu Palevol*, Vol. 14, pp 353-362.

———, G. et al., 2021, Dating of the Lower Pleistocene Vertebrate Site of Tsiotra Vryssi (Mygdonia Basin, Greece): Biochronology, Magnetostratigraphy, and Cosmogenic Radionuclides, *Quaternary*, Vol. 4, No. 1.

Mantzouka, D. et al., 2019, Two fossil conifer species from the Neogene of Alonissos Island (Iliodroma, Greece), *GEODIVERSITAS*, Vol. 41 (3).

Pfeiffer, B., 2023, "Vultures all the way down," Boston Globe, March 19, 2023.

Roussiakis, S. et al., 2019, Pikermi: a classical European fossil mammal geotype in the spotlight, Geological Heritage in Europe, *European Geologist Jour.* 48, Nov. 14, 2019.

Smith, B. J. et al., 2002, Sources of non-glacial, loess-size quartz silt and the origins of "desert loess," *Earth Science Reviews*, Vol. 59, Iss. 1-4, pp 1-26.

Sommer, R. S. & Nadachowski, A., 2006, Glacial refugia of mammals in Europe: evidence form fossil records, *Mammal Rev.*, Volume 36, No. 4, pp 251-265.

Taberlet, P. et al., 1998, Comparative phylogeography and postglacial colonization routes in Europe, *Molecular Ecology*, Vol. 7, pp. 453-464.

Theodorou, G. E. et al., 2010, Mammalian Remains from a New Site Near the Classical Locality of Pikermi (Attica, Greece) Proceedings of the XIX CBGA Congress, Thessaloniki, Greece, Special Vol. 99, pp 109-119.

Tourloukis et al., 2018, Magnetostratigraphic and chronostratigraphic constraints on the Marathousa 1 Lower Palaeolithic site and the Middle Pleistocene deposits of the Megalopolis basin, Greece, *Quaternary International*, 497: 154-169.

Trofimov, B. A., 1962, Osnovy paleontologii: Mlekopitaiushchie. Mosco-Leningrad, from: *The Great Soviet Encyclopedia*, 3rd Edition (1970-1979). © 2010 The Gale Group, Inc. via the freedictionary.com.

Tsoukala, E. et al., 2011, Elephas antiquus in Greece: New finds and a reappraisal of older material (Mammalia, Proboscidea, Elephantidae) *Quaternary International*, Vol. 245, pp 339-349.

Tzedakis, C., 2004, The Balkans as Prime Glacial Refugial Territory of European Temperate Trees, Chapter in book: *Balkan Biodiversity* (pp 49-68).

Woodward, A. S., 1901, On the bone beds of Pikermi, Attica and on similar deposits in northern Euboea, *Geological Magazine*, Vol. 8, Iss. 11, pp 482-486.

CHAPTER TEN:

ATHENS AND THE ACROPOLIS

THE OLD EGYPTIAN PRIEST, AS WE RECALL, FOCUSES particularly on the Acropolis, and that is where we will now turn to finish our examination of the geology of Greece. We touched briefly on the Athens Basin when we examined the Mesogea Basin to its east, and we'll see that the Athens Basin is much like all the others; the big difference is that here we have what seems, and indeed, almost purports to be, something approaching an eye-witness report of its geological history, from the person of the Egyptian priest. The confident way he speaks of ancient Athens gives one the impression that he actually saw the Acropolis for himself at some point, and he possibly did, and thus knows the present condition of it, and its surroundings.

Now, to remind ourselves of what the priest said of anything and everything that could possibly relate to the geology of the Acropolis:

> "The layout of the city in those days was as follows. The Acropolis was different from what it is now. Today it is quite bare of soil that was all washed away in one appalling night of flood, by a combination of earthquakes and the third terrible deluge before that of Deucalion. Before that, in earlier days, it extended to the Eridanus and Ilissus,

it included the Pnyx and was bounded on the opposite side by the Lycabettos; it was covered with soil and for the most part level. Outside, on its immediate slopes, lived the craftsmen and agricultural workers who worked in the neighborhood." . . . "There was a single spring in the area of the Acropolis, which was subsequently choked by the earthquakes and survives as only a few small trickles in the vicinity; in those days there was ample supply of good water both in winter and summer . . ."

The priest relates how back in the days of Atlantis, the Acropolis wasn't like it is now, as shown in fig. 10-1. In those days, it was much more extensive, deeply covered with soil (which we have to take to mean any kind of loose material, such as gravel, sand, loam, etc.) and it was level-ish on top.

Fig. 10-1: The bare and steep-sided Acropolis.
Wikimedia, CC-BY-2.0 License, photo: Carole Raddato.

The present-day Acropolis is only flat on top because distinctively large retaining walls were built all round and the hollows inside filled up to provide the present surface on which the Parthenon is built. It was also much larger in area, incorporating Pnyx Hill within it, and Pnyx Hill is a few hundred meters to the lower left of the Acropolis in fig. 10-1, with carved limestone steps being cut out of the rock of it.

Hence the whole space between the Acropolis and the nub of Pnyx Hill was filled with material which, by implication, buried both hills also, and extended beyond Pnyx Hill, since the Acropolis "incorporated it" as the priest says. This is an awful lot of material to have been swept away, and, if the priest is to be believed, obviously no ordinary, everyday process was involved. There is, of course, no need to "believe" the priest; we need only examine the evidence. One thing is certain, however, the Acropolis as it is now, could not possibly have accommodated 20,000 soldiers on top or any farmers on its cliffsides.

Since the ancient Acropolis (supposedly) also extended to the rivers Eridanus and Ilisssus, and nearly to Mount Lycabettus in the almost opposite direction to Pnyx Hill, it must have been very extensive indeed. A good idea of its extent can be estimated from the image in fig. 10-2. In the foreground, but not shown, is the river Ilissus; the Acropolis is at top center with Pynx Hill to the lower left of it, hidden by trees, and in the distance to the right beyond the Acropolis is Mount Lycabettus, meaning the ancient Acropolis was rather massive compared to today's, all this area being a single hill, according to the priest's description. Judging by fig. 10-2, this "hill" would seem to have been more of an extensive plateau, covering the area and providing room for all the soldiers and farmers.

Fig. 10-2: View across Athens showing relationship of features mentioned in text, Acropolis in the middle foreground, with Pynx Hill to the lower left, Mt. Lycabettus is beyond to the right. Image: ID 1208865241 © Stock Photos, Acropolis—Athens, photo: Rawf8. Royalty-free Lic.

There is no question but that the Acropolis, as it is today, represents a heavily eroded remnant of a much larger mass of rock; its rough cliff-like sides appear to have been severely battered by some agent, and this agent must also have removed whatever loose (soil) covering was over it, along with whatever rock that agent eroded away. At the same time, there does not appear to be any erosion, or even weathering, to speak of going on at present. Mt. Lycabettus is in a similar condition.

Athens is in a fairly dry and warm area, with cool winters during which it receives some rain, but little to no frost, and some rain does seep down into cracks in the rock mass and apparently causes dissolution of the limestone in faults and cracks (Regueiro et al., 2014, p. 51). Dissolution of limestone, however, is often simply assumed, without necessarily checking to see just how soluble, or otherwise, any limestone rock might happen to be.

Fig. 10-3: The heavily eroded Acropolis with retaining walls around the top. The rocky sides are bare and little material is seen around the base, indicating that whatever eroded it is no longer operating. Image: ID 887914520 © Stock Photos, Acropolis Athens, photo, Biggunsband. Royalty-free Lic.

Weathering due to expansion and contraction is also not apparent. A glance at the slopes and base of the cliff reveal little in the way of loose debris that might be described as weathering product, and there are no rivers or streams coming off the Acropolis that might wash any material away. There are only some springs near the base of the hill, but these produce mere trickles.

Thus, if weathering has been ongoing for eons, where is all the product, the debris, etc.? Any glance at the minor amount of loose material lying around the base of the hill confirms this, and that the hill has been in its present condition for a long time, apart from a few odd chunks that fall now and again, which are utterly trivial in comparison to the vast amount previously removed. Human inter-ference has more to do with any ongoing deterioration than anything else, as old rainwater drainpipes rusting out were found to be respon-sible for a recent rockfall (ANA News, Greece). The photograph of fig. 10-3 gives a very clear image of the eroded remnant of rock the hill of the Acropolis now is.

The rocky core of the Acropolis might be called either a very large "tor" or an "inselberg," each of which typically consists of a bare rocky prominence, usually rough and heavily eroded, that proj-ects above the surface of a generally smooth-sided and more gently sloping hill (tor) or rises above a flat plain or gently sloping surface (inselberg). The lower part of a tor's hill-shape and surface is typically formed by a covering of loose material blanketing the flanks of the underlying bedrock form.

Tors are very common in the formerly glaciated regions of Northern Europe and America and have been studied since the early days of geology. Oddly enough, there is still no agreement among the geological community as to how they were formed, apart from the obvious one of weathering and eroding agents of *some* kind (Linton, 1955).

The Acropolis, therefore, also falls into the same category as its northerly counterparts, the "as yet unexplained." Therefore, some erosive agent, clearly not operating today, must have been responsi-ble for the current state of the Acropolis, and the erosion caused by that agent must be the last thing to have happened to it, regardless of what that agent was, when it operated, or for how long.

The diagram of fig. 10-4 is a geological cross-section of the Acropolis hill showing the rounded-off hill-top, the faulted nature of the hill and the retaining walls built up from the cliff-edges. The "cataclastic limestone" is a layer of broken-up rock (breccia) that forms the thrust plane along which the "upper limestone" was pushed over the lower "Athenian schist." While it has the form of a large tor, the upper part of the Acropolis is a remnant chunk of foreign rock that was pushed into place over the local rock and is thus technically termed a "klippe."

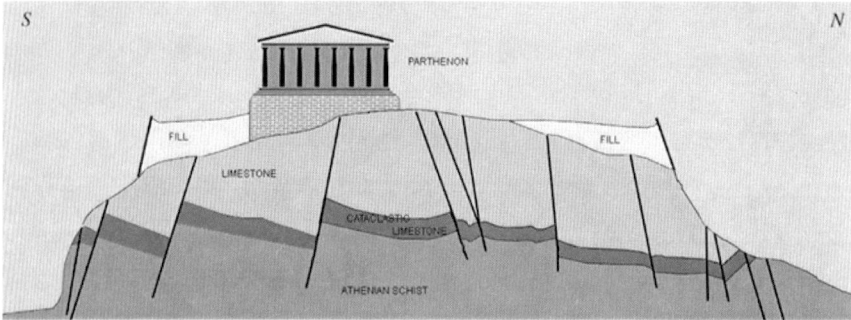

Fig. 10-4: Geological cross-section of the Acropolis. From Regueiro, Stamatakis & Laskaridis (2014), (adapted from Higgins & Higgins, 1976).

According to Regueiro, et al. (2014) in their monograph on its geology, the Acroplis klippe was formed when the limestone was thrust over the schist 30 my ago during the Alpine orogeny. That it was an "abrupt" movement, as the authors state, is demonstrated by the cataclastic layer at the base of the limestone. This layer is made up of upper limestone fragments (clasts) of all sizes, showing shearing and flowing textures, and results from being dragged, at some depth and at high temperatures, under the pressure of the overlying limestone layer, and in what is suggestive of a catastrophic manner (hence the "cata" prefix on cataclastic). Figure 10-5 shows a section of that cataclastic layer with large chunks of limestone, along with a multitude of smaller fragments in a reddish matrix of finely ground rock flour. The top of the layer is easily seen, with a rough stone wall built on it.

Fig. 10-5: Cataclastic rock (cataclastite) layer underlying the upper limestone of the Acropolis. The rounded clasts are of upper limestone and their shape implies rolling movement which is associated with the thrust planes. Photo: Callan Bentley, Geology of the Acropolis (Athens, Greece), January, 2015.

Now, the authors here do not specify exactly what they mean by "abrupt." However, slow, creeping movement at high heat and pressure will not fragment rock, and certainly not round the clasts as we see them in the photograph; it will grind it fine, (partially) melt or plasticize it and hence make it ductile. The typical, and theoretical, uniformitarian rates of movement, of a few cms a year, are far too low to cause fragmentation, and we do not see brecciation going on to any extent today at fault planes. Like much else, we see that it has occurred in the past, but, in true uniformitarian fashion, it is not now seen to be occurring.

That is not surprising since we see little, if any, movement on any fault planes today, apart from major ones such as the San Andreas in California, and this despite the millions of fractures we see all over the world. Therefore, we can be confident that the former movement on the thrust plane that put the Acropolis klippe into place occurred quickly, and continuously, over a considerable distance and area, and was far outside anything occurring at the present.

And, while I'd take the authors' estimated age of 30 my with my usual seasoning, I don't necessarily think that this "event" was very recent either. However, as we'll see later, the time period of the formation of the Alps is very much open to question. Whatever is actually the case here, our main interest is in what is lying on top of the solid bedrock of the country.

Unfortunately, Regueiro et al. (2014) do not give any information at all on what's called the "overburden," the loose materials of various kinds that lie over the bedrock, as such material does in most regions of the world. The authors do tell us, however, that the rocks in the general Athens area show severe weathering, intense folding, shearing and brecciation, along with extensional faulting, indicating that the area has suffered intense, and apparently violent, disturbance in the past, much like the rest of Greece and the Aegean.

The main body of the Acropolis limestone exhibits similar signs of fracture and deformation. The authors conclude their description of the origins and geology of the Acropolis itself with the following rather brusque, and entirely uninformative, statement: "After this process, erosion did the rest." And so, here again, we must take our duly directed leap of faith and fall back on uniformitarian indoctrination. However, this is the whole question; *what* accomplished the erosion? The answer to that is the very geological history of the Acropolis.

Athens receives about 15 or 16 inches of rain a year on average, gets very little frost, and rarely sees more than a dusting of snow. It does occasionally experience severe rainstorms and flash flooding, which do cause damage, but are, of course restricted to low-lying areas and are usually composed of muddy water with debris of various kinds. These floods do wash out roads and other structures but do little in the way of erosion of even the unconsolidated sediments in the Athens area, let alone the hard rocks.

These floods are not to be confused with the much more powerful and violent mountain torrents that sweep boulders and everything else down a mountain. No streams wash or wear the Acropolis and rain has no power to erode solid rock, nor has rain got much ability to dissolve the limestone of the Acropolis either since so little of it falls on Athens. Further, and unsurprisingly, there is little or no sign of frost action. There are, therefore, few "everyday causes" in operation here.

Thus, while the authors blithely say "erosion did the rest," they have obviously taken it for granted that the usual everyday process, working over the usual long ages, have rendered the Acropolis in its present condition. However, as noted, little to nothing in the way of erosion is going on at present, and has not for much of (long) recent history, since soil has obviously developed on the hill's flanks and they are reasonably covered with trees and other plant life. Hence, whatever it was that eroded the Acropolis was something entirely different to the usual "everyday processes."

While the authors say nothing of the surface overburden of unconsolidated (and therefore Ice Age and recent) sediments in the area, they do provide a geological map of the Athens region showing the areas where the upper limestone and the Athenian schist outcrop. The map also identifies the type and areal extent of the different materials that comprise that surface overburden, which lies on the Athenian schist, which itself underlies the entire Athens area. Fig. 10-6 is a geological map, one of the sheets of the Geological Map of Greece, that details the Athens area and shows both solid rock and sedimentary overburden.

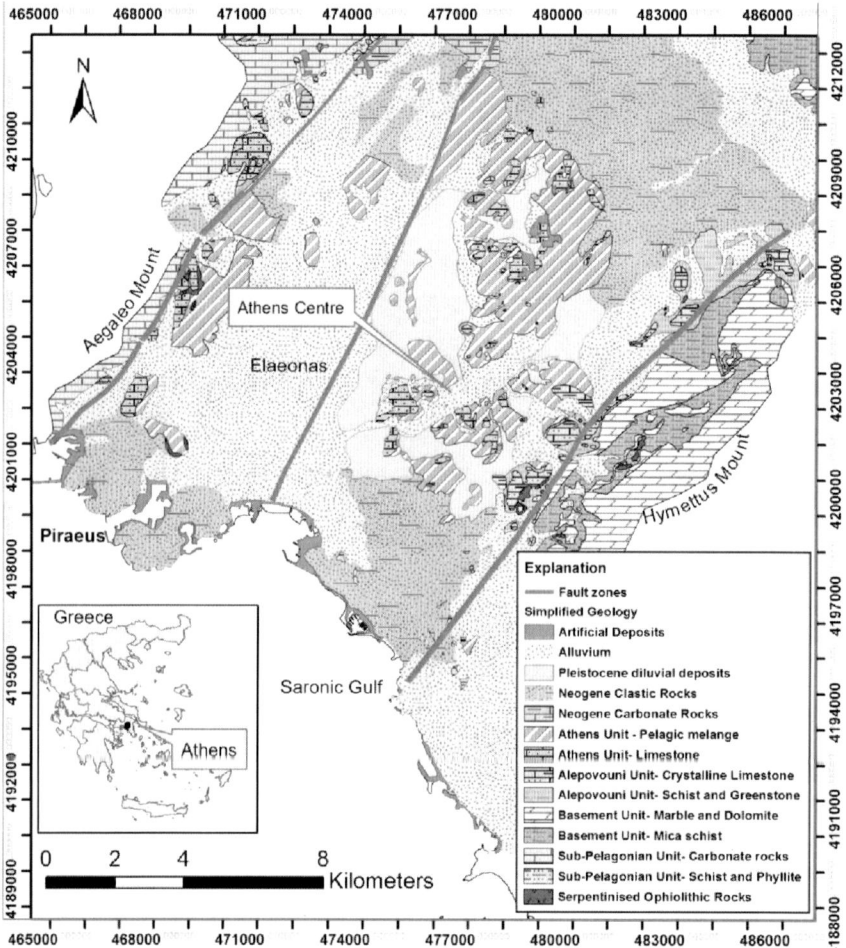

Fig. 10-6: Geological map of the Athens area showing the various rock types and overlying sediments. The key in the lower right shows that the basin is filled with clastic rocks, Pleistocene *diluvial* deposits and recent alluvial deposits dating from the end of the Ice Age and with the Acropolis hill surrounded by the same. Image: Argyraki & Kelepertzis. (2014).

The city of Athens is situated in the Kifissos river basin, the Kifissos being the biggest river that drains the Athenian plain and of which the two rivers mentioned earlier, the Ilissus was a former tributary, until human interference diverted it, while the Eridanus has dried up, leaving an empty bed. The plain contains a number of hills, of which one is the Acropolis, while others are represented by the diagonally striped areas in the map of fig. 10-6, and shown in the photo of fig. 10-2. The basin itself is similar to the other grabens discussed earlier, and contains sediments of the usual varieties, clays, sands, gravels, etc.

Interestingly, while the geological map shown above does not describe the nature of the sediments, the key to the map (lower right) identifies "Pleistocene *diluvial* deposits," i.e., flood deposits, surrounding the Acropolis.

As we have seen, the condition of the Acropolis as it is now is by no means as straightforward and as easy to explain as convention has it. Like all the basins in Greece, the Kifissos too is filled with water-transported sediment. And, as noted elsewhere, the term *diluvial* is put to surprisingly regular use by presumably uniformitarian geologists, though, of course, in this context, or in any use of the term, orthodox geologists simply mean an ordinary, or extraordinary flood, such as we might get for some reason every 100 or 1,000 years or so.

The map of fig. 10-6 shows the Acropolis and surrounding areas covered with a blanket of water-transported sediment that is draped over the surfaces and flanks of these hills. Water does not flow uphill, and the rivers presently flowing in the Athens Basin could, obviously, in flood or not, have had nothing to do with the deposition of that Pleistocene sediment up on the hillsides (or filling the basin from one side to the other, or end to end).

Some much more powerful and widely acting agent must, therefore, have been responsible, as it appears to have been everywhere else. (This, as the reader will likely recognize, is exactly the same kind of material that Bald and Buckland were referring to in the epilogue to volume 1, "old alluvium" or "diluvium," and up on hillsides just like their own.)

I have not been able to locate much information on the deeper sediments of the Athens Basin, at least not in English, but E. D. Chiotis (2016) tells us that there is a considerable amount of recent alluvial sediment, up to 60 m (180 ft.) thick. This is a lot of sediment for some agent to deposit recently, and it could not have been the present-day rivers since they're far too feeble. Chiotis reports that it consists of, as usual, a number of units relating to sea level rise, or transgressions and regressions, i.e., the sea invaded the land and retreated; and this occurred very slowly of course.

When such a thing happens quickly, we call it a tsunami. However, like many other writers on Greece, he identifies some high water-level around 5,000 years ago. Considering what we heard from Zangger et al., it is by now obvious that something notable occurred in Greece at about this time, but, since it was only 5,000 years ago, it had nothing to do with Atlantis, but it may well have been responsible for Chiotis' recent alluvium.

To investigate the geological history of the Acropolis and the Athens Basin, we will have to look at the wider region, including adjacent basins and mountains, etc. Fortunately, because of the Pikermi fossils, the Mesogea Basin to the east of Athens has been very well studied, as we've already seen. And, therefore, we already know that this basin has been subjected to massive floods sweeping back-and-forth through it—and well-beyond, into the Aegean.

As shown in fig. 10-7, this basin is one of three adjacent basins, comprising an arrangement, or complex, similar to those we've seen in the Aegean and in northern Greece, such as the Amorgos-Astypalea or the Florina-Ptolemais basins respectively.

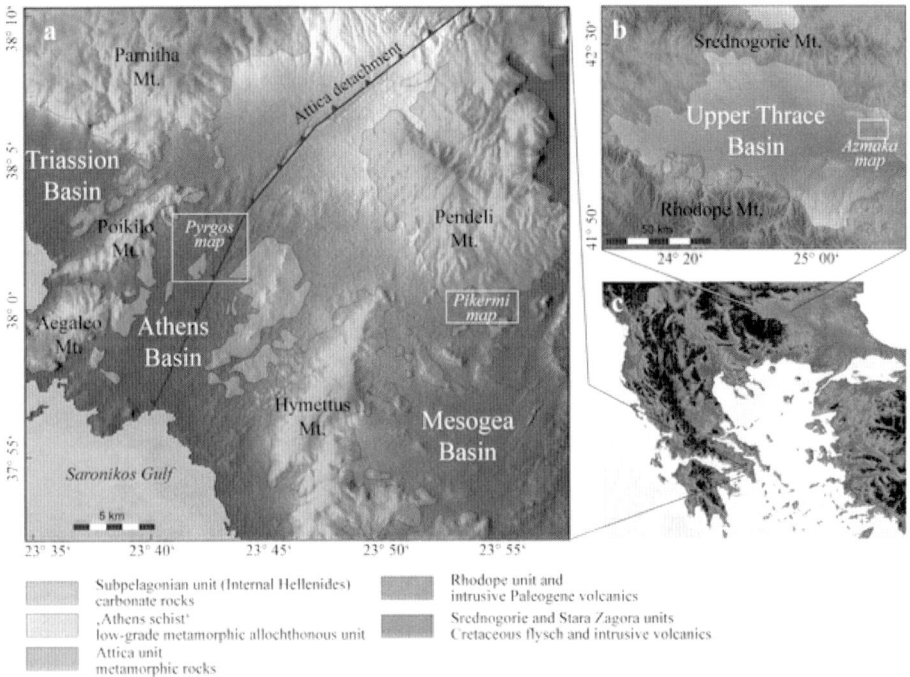

Fig. 10-7: Map showing triple basin complex typical of those examined so far, along with the surrounding horsts. Important here are the different rock types that make up these intervening horsts. To the west the Athens basin is bounded by carbonate rocks of Parnithia, Aegaleo and Poikilo Mts., while to the east lie the metamorphic rocks of Hymettus and Pendeli Mts. (also known as Penteli Mt. or Pentelikon Massif). These rocks are easily identified, and an analysis of the sediments in the basin can reveal their origin. This, along with the size and shape of fragments, and other features, can be used to estimate the mode of sediment transport and the magnitude of the forces involved. Image: Böhme et al., 2017.

As can be seen, the basins are separated by ridges (the usual horsts) while the area in the center of the Athens Basin and north-west of Hymettus Mtn. comprises the hills of Athens itself, including the Acropolis. The different rock types are prominently shown, limestones and other carbonate rocks to the west, with metamorphics to the east. The square area in the Athens Basin labelled *Pyrgos map* is another location where Pikermi fossils were found, though some variation or differences with Pikermi are claimed (Böhme et al., 2017), which wouldn't surprise me and is of little significance.

While the writers so far consulted were focused primarily on the Tertiary fossils found in particular sedimentary formations of the Mesogea basin, Mposkos et al. (2007) studied it in an effort to elucidate its more recent (within the last 5 my) development, its "geodynamic evolution" as they put it. This involved a combined study of the structure of the basin and the sediments it contains. They agree with practically every other writer on the basins of Greece that the Mesogea Basin also contains the usual three sedimentary series. The basin has the same origin as the many others, faulting, extension and subsidence.

The study of its Miocene (15 my) to Pliocene (2 my) evolution is based to a great extent on the nature and origin of its sedimentary infill. The study begins by telling us that: "tectonic movements associated with basin formation have continued until recently" (Mposkos et al., 2007) in which case, we can conclude that, like everywhere else in Greece, something notable occurred here not long ago. "Continued until recently . . ." is simply the usual uniformitarian "ongoing process" assumption applied to everything.

The reader will by now have noticed that all tectonic activity such as earthquakes and plate movements are always presented as "ongoing," but little else is, like Ice Ages or lake-filling for example. I find it rather curious and somewhat amusing that the only processes and agents that are so ongoing are those that are deep in the earth and can't be seen, perfect candidates for "ongoing leaps of faith." Fig. 10-8 is a simplified tectonic/geological map of the basins and mountain formations of Attica.

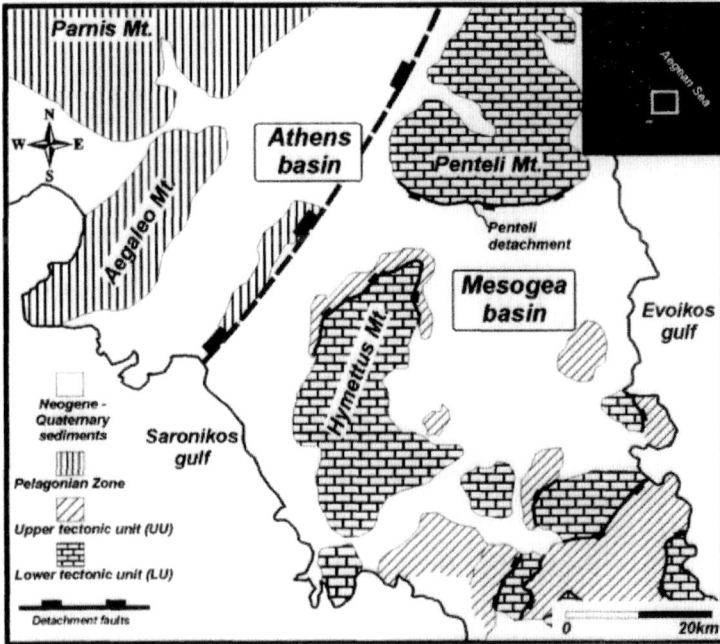

Fig. 10-8: Map showing Mesogea and Athens Basins and surrounding mountains. The geology of the mountains is simplified to rock type only. As can be seen, the Mesogea Basin is surrounded by mountains of the Lower tectonic unit, Mt. Penteli and Mt. Hymettus (block-shaded) while the Athens Basin is walled on the east by the same rocks and on the west by rocks of the Pelagonian Unit (vertical shaded). Some Upper Tectonic Unit rocks crop out in both basins (diagonal shading). Image: Mposkos et al. (2007).

As mentioned, the infill of the basin has been divided into three distinct series. Many workers over the last century, and up to quite recently, in fact, have described these three sedimentary series, which consist of ". . . laterally changing sandstones, mudstones, conglomerates and limestones . . .," as "diluvial" sediments, by which they simply meant flood-derived, while modern writers usually refer to them only as "fluvio-terrestrial" and "lacustrine."

However, we've seen enough by now to know that any mention of the word *conglomerates* must imply severe flooding of some kind, and conglomerates on land are nothing if not "fluvio-terrestrial." The muds and limestones may or may not actually be "lacustrine" and by now I've reason to doubt, and the conglomerates, and everything in between, certainly aren't lacustrine, raising the obvious question as to whether the muds and limestone are.

Having studied the clasts in the conglomerates and sandstones, the authors note that a large proportion: ". . . originate from sedimentary and ophiolitic rocks that are notably absent in the mountain ranges now bordering the basin" (Mposkos et al., 2007). Based on the rock type, these clasts seem to have been derived from the Pelagonian zone to the west of Athens, where such rocks are common on Mt. Aegaleo and Mt. Parnis (Parnithia Mt.) 40 km (25 miles) away, as well as on the hills of Athens in the center of the Athens Basin, and do not occur anywhere else in the area, as can be seen in figs. 10-7 and 10-8, above.

The authors do tell us that the character of the clasts, which means their shape and size, suggests ". . . either material transport over long distances or smooth relief." Since they come from the mountains to the west of Athens, it was certainly long-distance transport, whatever about the means.

The suggestion regarding "smooth relief" is a very old one; in fact, it's a very "venerable" one, and from the pen of one of the very founding fathers of uniformitarianism. It is something of an "escape chute" often resorted to in situations like this, where there is no apparent *uniformitarian* way to transport material from a distant source to where it is found, particularly when a large valley intervenes, and the writer can't call upon the Ice Age to do the work.

Hence, a "smooth inclined plane" is invoked as having formerly extended from source to deposit, thereby (theoretically at least) allowing a supposedly ordinary stream to supposedly account for the transport of the clasts all that 40 km distance.

Later, all of the enormous, intervening mass of rock is somehow eroded away, to leave a valley or basin in between, the processes involved in such an enormous amount of erosion and transportation never actually being identified. This idea was first proffered by John Playfair in his *Illustrations of the Huttonian Theory* to account for boulder transport (Playfair, 1802, p. 385), and was based on a mention by Hutton regarding glacier transport. Playfair suggested transport on top of a glacier (as the inclined plane) to explain Alpine boulders high up on the Jura mountains, on the other side of a wide valley, a situation similar to what we have here in the Mesogea basin.

However, apart from the fact that we know for certain that no intervening rock mass ever pre-existed in this basin, given that the structural interpretation of both basins and mountains is a horst and graben system, such a shallow inclined plane, stretched out over 25 miles (40 km) with a river flowing on top of it couldn't explain the conglomerate in any case. Ordinary rivers, such as are implied here, flowing slowly on the near-plain that this "intervening inclined plane" would have constituted, by physical imperative, can't move heavy material, and the authors tell us that the clasts in the different conglomerate layers vary in size from 2 or 3 cm to 15 cm (or 1–6 ins.) and no ordinary river can move that size material.

A meltwater or rainstorm flood coming down a mountain valley certainly will, but that is not what has been proposed here, because it can't be, given that there's less evidence for mountains *in the basin* than there is for a shallow slope extending all the way from Aegaleo Mt. in the west to the Mesogea basin in the east, and there's none for that.

The authors very studiously avoid mentioning the Pelagonian Zone to the west of Athens as the obvious source of much of the material in the Mesogea basin. They instead restrict themselves to a very generalistic identification of the rock types, i.e., limestones, sandstones, mudstones, metamorphics, etc., and hence avoid directly identifying the Pelagonian Zone, or the Athens hills, the only sources of this material.

It is obvious that a rather simple explanation would be that a flood swept in from a northwest direction and overran the mountains to the west of Athens, tore masses of debris from them, which it then swept across the Athens Basin, up and over the hills of Athens itself, clearing everything from them and battering the Acropolis down to a nub, and by this means and agency rendered the Acropolis as we now find it.

This flood then swept on between and over the Hymettus and Penteli Mts. and into the Mesogea Basin, where it naturally slowed down, spread out, and deposited much of its load, all well-rounded from long-distance travel, as the authors inform us. As we saw in the last chapter, this flood then swept on into the Aegean, came to a halt, and swept back to bring all those barnacles, etc., into the basin. From a glance at the map of fig. 10-7, the track of this proposed flood can easily be envisaged, while the "torrential floods" from Penteli Mt. are automatically explained.

The authors also report that some of the material of Series I and II derives from Penteli Mt., based on the type of metamorphic rock. This, of course, means it could also have derived from Hymettus Mt., which is more in line with a flood from the northwest. Whichever is the case, material from the distant west, mixed with material from nearby, is consistent with a major flood having swept over the area.

The upper unit, Series III, lying above an erosion surface, derives, according to Mposkos et al., from Penteli Mt to the north, based on the usual analysis of its clasts and fragments but, as stated, it could also have come from Hymettus Mt. to the west. An erosional "event," marked by an erosional surface at the top of Series II, is clearly implied prior to deposition of Series III. We have met such phenomena before, and this sounds like the usual "erosion surface" underlying a typical flood deposit, just like everywhere else.

The change in provenance of the clasts in the sediment is such that the authors consider it "remarkable" and tell us that the clasts are angular, implying short transport distance, and that boulders up to 1 m (3 ft.) in diameter are present, in what are the usual "fluvio-terrestrial" conglomerates. In other words, the direction of transport changed from the northwest to the north, while the authors explain this remarkable evidence by rousing the ghost of W. M. Davis and invoking his "uplift" as their "relief formation" to account for the increased erosion, though, as usual, no evidence for the uplift is provided.

They have no real explanation for either the uplift or the boulders because they are not allowed to invoke the obvious flood coming from the north, over Penteli Mt., though I consider Hymettus Mt. to be the more likely source. However, they do agree with other writers we've consulted (Böhme; Dermitzakis) and, based on all that we've seen in Greece thus far, a major back-and-forth flood would appear to be the correct and only possible explanation.

Earlier, I pointed out the very uneven and channeled nature of the southern slope of Penteli Mt. as seen in fig. 9-4. This is due, according to Mposkos et al. (2007) to large blocks of marble bedrock having broken off the mountain, which then slid down and out onto the sedimentary fill of the basin. These blocks are referred to as "gravity-sliding" blocks which broke off from the mountain along "detachment-faults."

The blocks lie above Series II and within Series III, meaning that this "detachment" of huge chunks of the mountain occurred very recently, which is to say during the deposition of the (latest) Series III sediments, since the blocks are described as being both *on* it and *in* it. Thus, it would appear that this event occurred since the Ice Age, and likely at its end, like much else.

Such blocks are also reported from isolated hills within the basin toward the north side (Mposkos et al., 2007). Huge chunks of mountain breaking off and sliding down into the basin suggests, to me at least, that this whole area, mountains, basins and all, has very recently been violently shaken, and, according to the evidence, shaken to bits, in fact.

While everything I've said, based on evidence provided by academic geologists, goes against uniformitarianism, we nevertheless now also have a more reasonable explanation for the condition of the Acropolis as we find it, no matter how unreasonable it sounds. It wasn't, simply and innocently, that "erosion did the rest;" a gigantic flood came over the mountains from the west-northwest, tore off and swept up enormous masses of rocks and debris, and then overran the Acropolis, battering it to pieces, removing everything on it, and rounding it off, leaving material draped over its lower flanks. And that flood swept back and forth three times, as we saw earlier when we examined the Pikermi fauna.

And, in support of that statement, and apart from the geological evidence we've seen, I simply claim that that's what looks like happened to it. The Acropolis, to me, does look as if a rubble-heavy flood swept over it multiple times, stripping it bare and eroding it away, especially on its western and eastern sides. And as that flood slowed down, it deposited a blanket of sediment around the lower flanks of the hill, leaving it just like a tor as seen in formerly "glaciated" districts.

Having examined the geology of Greece and the Aegean from various perspectives, we find the explanations from orthodox geology to be seriously wanting. Everywhere we looked, the evidence did not seem to comply with the theories of that evidence's origin, formation or development, or with uniformitarian geology in general. We have gone into some detail, and while that old Egyptian priest painted his canvas with broad strokes, we find the detailed picture of the geologists to be quite compatible with the priest's rough sketch.

The big differences are the usual ones—time and energy. Geology, as usual, uses the traditional uniformitarian low energy over a long time, while the priest's version is the standard catastrophist one, high energy, short time, and which would appear to be the correct one, from the point of view of the evidence and physics.

We have certainly seen evidence to indicate lots of high energy; those big boulders take a bit of effort to move and no amount of time in the world will move them without the necessary *minimum* energy, as the laws of physics tell us, and this minimum is simply the *force threshold* which the applied force must exceed to move that rock.

And the reason mainstream geologists have such trouble explaining so many phenomena, in any kind of convincing way, is because they have been convinced, since the glory days of the British empire, that anything is possible if only one considers time and energy the same thing, has an infinite store of patience, a mind that does not deal with practicalities, and never heard of the laws of physics.

Having gone through most of the more recent geology of Greece, what orthodoxy might consider the last 20 my or so, we have met with issue after issue whose "explanation" does not seem to measure up to the requirements of any particular case. We "know" that the Ice Age was the last major "event" to befall this planet, supposedly, but we've already seen in Greece that it was not. Many workers, Zangger being prominent, report a more recent occurrence, which some have termed an alluviation event, or flooding event, or some such, with words like *paroxysm, megaflood, seismic event,* etc., being liberally used.

However, it would appear that the geology of Greece holds another mystery for us. And it lies right on top of everything else, in the form of the red soils that comprise much of the better agricultural land of Greece, and which are prevalent over the entire Mediterranean Basin as well as being found in Africa, the Middle East, Asia, and indeed, many areas of the world. This red soil is most often referred to by its Italian name, *Terra Rossa,* red earth, and this is the same red, earthy gravels that Captain Spratt described on Crete.

This mystery really has two parts; the first is its origin, or genesis, while the second is its distribution, the second rendering any solution to the first rather difficult. The main problem is that it has an essentially global, though irregular, distribution, in different geological contexts, and in connection with many different rock types, while at the same time, its composition does not vary much over regionally extensive areas.

This has caused a problem with regard to its genesis, because, as usual, the instinctively uniformitarian explanation would have it that the red soil must derive from some process or other, currently ongoing of course, at every location where it is found, and since it is the same everywhere, it *must be* the same process everywhere.

Chapter References

Argyraki, A. & Kelepertzis, E., 2014, Urban soil geochemistry in Athens, Greece: The importance of local geology in controlling the distribution of potentially harmful trace elements, *Science of The Total Environment*, Vol. 482-483, pp 366-377.

Bentley, Callan, 2015, Geology of the Acropolis (Athens, Greece). After Regueiro et al., 2014, posted to Mountain Beltway blog.

Chiotis, E. D., 2016, Landscape evolution in the Kifissos floodplain, Chapter 23, *Proceedings of the 6th symposium of the Hellenic Society for Archaeometry.*

Linton, David L., 1955, The Problem of Tors, *The Geographical Journal*, Vol. 121, No. 4, p. 470.

Mposkos, E. et al, 2007, Late- and Post-Miocene Geodynamic Evolution of the Mesogea Basin (East Attica, Greece): Constraints from Sediment Petrography and Structures, *Bull. Geol. Soc. Greece*, Vol. XXXVII, Proceedings of the 11th International Congress, Athens, May, 2007.

Playfair, John, 1802, *Illustrations of the Huttonian Theory of the Earth*, Facsimile reprint, Dover Publications, Inc., New York, 1964.

Regueiro, M, Stamatakis, M, Laskaridis, K., 2014, The Geology of the Acropolis (Athens, Greece), *European Geologist*, Vol. 38, Nov. 2014 pp 47-48.

CHAPTER ELEVEN:

THE RED SOILS OF GREECE AND THE MEDITERRANEAN

WE HAVE, OF COURSE, MET RED SOILS IN GREECE already, in many places and at various levels in the geological record. We recall the extensive, thick beds of red silt we met at Pikermi, fig. 9-9, and the alluvial fans of the Xirolaki stream, shown in fig. 6-3. As far as Pikermi went, the source proffered was the Sahara, and this same source is commonly claimed as having something to do with the red surface soils of the present also. Obviously, no one has read Velikovsky lately.

However, the traditional explanation for the red soil, or just red silt, is that it results from the dissolution of limestone, and I have to presume that this is because it was first encountered directly on limestone, in the Mediterranean region (Italy) by 19th century European geologists, who simply opted for the uniformitarian explanation.

All over the Mediterranean basin, the red soil is found lying on limestone, much of it karstified by water dissolution. My conclusion is that, following the Lyellian practice of ascribing everything to some ongoing process or another, these geologists saw the red soil directly on the limestone and simply associated them genetically.

While by now, few geologists hold to the limestone theory simply because it is, among other things, a chemical impossibility, the second most favored explanation, at least for the Mediterranean region, is Saharan dust, just like at Pikermi. However, some geologists still cling to the limestone origin.

The chemical formula of limestone is $CaCO_3$, a calcium, a carbon and three oxygens. However, to get a red color one must have iron (Fe) or aluminum (Al) and usually both are found as silicates, and there is obviously none in $CaCO_3$. There may be some iron in limestone, as a constituent in whatever clays or other impurities may have been incorporated during its formation, but it is always a miniscule amount, usually less than one per cent.

One example is provided by Boero & Schwertmann (1989) who did a study of 48 terra rossa samples from various locations around the world, with a focus on the iron and aluminum mineralogy, haematite and goethite being types of iron mineral, and the samples were analyzed for Al substitution of the Fe. They found that the chemistry varied to a "rather limited extent" on a *global* level (author's italics).

Their interpretation of this is that the terra rossa forms in a rather specific environment, which is similar to the Mediterranean one, and it forms from the dissolution of limestone, while claiming that there's more than enough iron in the limestone to account for the terra rossa.

While there is usually some iron in limestone, as I said, it is not always present and is typically only a very minor constituent. They reject, without giving a reason, all arguments about insufficient iron, or a contribution from wind-blown (aeolian) dust. The fact that terra rossa is found lying on all manner of other rock types, as well as on loose sands and gravels and various other deposits, does not seem to have registered with these writers. Also, the significance of their own finding, of a general *global* similarity in composition of the terra rossa, equally seems to have been lost on them.

Fig 11-1: The famous Kokkinopilos terra rossa deposit in northwest Greece, which appears to be simply a massive deposit of red soil, considered to be similar to the surface of Mars, and entirely belying the notion of a limestone origin. Image: ID 419361459 © Adobe Stock Photo, photo by Yiannis Mantis, Royalty-free Lic.

Olson et al. (1980) working in Indiana, U.S.A., compared the thickness of the terra rossa with the composition of the limestone and found that there was nowhere near enough limestone to account for the quantity of terra rossa. They also found mineralogical differences, and a study of the soil indicated that it was the result of erosion of clastic rocks. They therefore reject the idea that it is a residual product of limestone weathering or erosion.

Clastic rocks here mean sandstones or maybe conglomerates, typically made up of fragments of granitic or other quartz-bearing rocks, which means that the terra rossa could possibly have come directly from the weathering or erosion of granitic-type rocks with no need for an intervening clastic-rock phase. The actual erosion process is not suggested by these writers, but we already know that it is very difficult to produce silt-sized particles from solid rock.

The important point is that the rock had to contain iron and/or aluminum, usually red in color, as in red sandstone. However, that just pushes the problem back a level, the source of the iron in red sandstone still hasn't been identified.

Geologist Michael May of Western Kansas University, in an informational article on his research, considers that the origin of terra rossa is not dissolution of limestone, for the same mis-matched volumetric reasons as stated by Olson et al., and considers instead that iron from elsewhere is replacing the calcium. As he says: "It doesn't make sense geochemically to be able to take a rock that is about ninety-nine per cent calcium, carbon and oxygen, and magically get all that iron by simple dissolution," which seems obvious to me.

His point is that the real mystery is where all the iron came from, and he surmises that it may just be "dust in the wind" just like we saw at Pikermi, and just as so many other geologists consider the red soil as aeolian in origin. Dust in the wind is all well and good, but from where? The answer to this question is definitely not "blowing in the wind," and certainly not from the direction of Lyell.

May collaborated with Enrique Merino and others at Indiana University, which, as we recall, is where we first met Dorothy Vitaliano. We find that when we consult Merino, he is of the same mind as May and rejects the dissolution solution, instead considering that the red soil's major chemical elements of Al, Fe, and Si probably came from dissolved aeolian dust, and these replaced calcium in the limestone.

No explanation is offered as to how the dust was "dissolved" or how it could possibly have replaced the calcium in the limestone, or why, and this suggestion just seems to indicate a fixation, on Merino's part, on the limestone origin, the better to toe the conservative line. There is again, however, no mention as to the origin of the dust (Merino & Banerjee, 2008). When we go to back to the Mediterranean Basin, we find that aeolian dust is becoming the new, and generally accepted, explanation for the soil everywhere.

Fig. 11-2: Terra Rossa as it's often seen, lying above limestone and filling cracks and voids, etc. Here the soil seems to have a lot of pebbles through it and is reminiscent of Spratt's "red earthy gravels." As May and Merino state, it is very difficult to see how so much red soil could be built up on top of limestone just from the dissolution of the underlying rock, not to mention how that dissolution could possibly occur. Image: Adobe Stock ID 205049940. Photo: Matauw. Royalty-free Lic.

In Greece, Yassoglou et al. (1997) describe the red soils as of two types, those that formed in place (autochthonous) and those that were transported (allochthonous). Those soils that formed in place are assumed to have had an iron-rich parental material, (by necessity) and are found on limestone as well as on igneous rocks, gneiss and other metamorphics, like marble and calcareous mica schists, and, of course, the mica schists that Spratt mentioned. However, if they're found on limestone and limestone doesn't have iron in it, how could the red soil have formed in place from iron-rich parental material in this particular case?

The allochthonous soils are described as widespread on both Tertiary (rock) surfaces, as well as on Pleistocene (gravel) surfaces, and are interbedded with marls and conglomerates, so there is no doubt about transportation, as Spratt also described. No mention is made of aeolian dust, or other origin for these Greek soils.

By 2010, however, and as we saw at Pikermi (Böhme et al. 2017) the Sahara was becoming the favored source. Muhs et al. (2010) determined, based on chemical analysis, that dust from the Sahara and Sahel regions played an important role in the origin of the soils of the Mediterranean region, even though their study mainly focused on Mallorca and Sardinia.

The similarity in color, apparently, also served as something of a "hint," which it would, obviously enough. In any case, they state that: "African dust explains the origin of the 'terra rossa' soils in the Mediterranean region on top of mother carbonate rock." Well, actually, no it doesn't. If the silt came from Africa, what is the basis for the mention of "mother carbonate rock?" That African silt is a motherless immigrant, not native-born to a limestone mother.

Having tested the chemical signatures of the rocks and the dust, and found them different, they concluded that the soil did not derive from the rock, and must thus have had an external source. They did not, however, attempt to explain where all that red dust of the Sahara and Sahel came from in the first place, which, of course, means that they explain nothing at all, and simply push the problem over to Africa.

The fact of the matter underlying all of this is that the red dust of the Sahara is as unexplained as all the other terra rossas and red earthy gravels and red silts and red clays to be found more or less everywhere, globally.

In an attempt to organize or rationalize the whole issue of the red Mediterranean soils and their genesis, Fedoroff & Courty (2013) present an overview of the contributions by various soil scientists, and geologists, over the last 100 years or so. They describe the various circumstances in which red soils are found, on limestone and other rocks, and also on Pleistocene deposits, as well as, significantly, buried in alluvial fans, intercalated with aeolian dusts, and stratified into layers, many of them gravelly. They point out that uniformitarianism, as historically applied to red soil formation, assumes growth in place (as always), and steady development to equilibrium (little by little, as usual).

However, the soils show variations (anomalies as they're here called) that necessitated the introduction of "subsidiary concepts" which is to say: as usual, invented processes and agents must be drafted in.

The problem, of course, is that the red soils show little evidence of the gradual, linear development uniformitarianism demands, which is no surprise to us at this point, because practically everywhere else we've looked, we've seen the same failure of Lyellian theory to account for the evidence. As Fedoroff & Courty report, these soils display abrupt changes in composition, and variations in their chemical makeup, in the absence of any evidence of climate change.

These changes are instead supposedly due to soil internal evolution (again, it seems, that great miracle-worker, "evolution," must be brought into it). This solution was no solution, so other ideas were proposed, i.e., geomorphic change, in that the relevant geomorphic processes are "periodic," which would seem to contradict uniformitarianism.

Another idea was a theory called "biorhexistasy" which is defined as the alternation of periods of soil formation followed by periods of erosion (Ehhart, 1956, cited in Federoff & Courty), and which sounds like yet another invented process.

Again, alternating periods of erosion and soil formation reflecting "abrupt changes," as mentioned above, necessarily bespeak rapid changes in conditions, which, again, are entirely contrary to the sacred tenets of uniformitarianism. Though at the same time one wonders how periods of simple erosion followed by periods of formation could lead to the creation of red soils in particular. Where's the iron or aluminum?

We can see the red soils just lying there with no sign of erosion or any formation of new soil going on, so what, one wonders, is the source of this seemingly futile suggestion, especially given there's no evidence of anything from the soils themselves.

The biggest mystery, by far, concerning the red soils is not actually their genesis; the idea that the positively vast quantities of red soils, to be found the world over, could be formed by the in-place weathering of underlying rock, or anything else, is unsupportable. These red soils attain thicknesses of hundreds of feet or meters in many places. It is clear that, as always happens with weathering processes, the weathering products will build up in place until such time as they eventually, and always, insulate the rock from further weathering.

We can verify this anywhere we do actually see weathering going on, like in the frost zone of a mountain; once the target is buried in the debris produced, the weathering agents, understandably, can no longer impact that target, and thus all weathering must cease, and this is what we see. Weathering agents are few in number, and as far as chemically weathering (corroding) rock is concerned, water is a prerequisite to sustain the chemical reaction. Therefore, as soon as the rock is buried in a few feet of detritus, that rock is isolated, and weathering ceases.

Another major mystery of the red soil, which includes many varieties, is their global extent and the fact that they lie more or less at the surface. Paepe et al. (2004) consider that the red soils of Greece, and the general Mediterranean, have analogues in other parts of the world, such as the loess areas of Northern Europe, Russia, China, the Americas, and the subtropical to tropical regions of Africa and Asia.

As far as Greece itself goes, they report multiple sequences of paleosols interspersed with fluvial sands, gravels and supposed aeolian deposits (loess), showing that transportation has more to do with the emplacement of the red soil than does genesis in place. This, again, is in line with what Spratt (1865) reported, from a time when Lyell's uniformitarianism hadn't yet so thoroughly infected the geological community.

The multiple layering of the beds studied by Paepe et al. (2004) led them to conclude that these sequences represented the climatic conditions and changes of the Pleistocene, and because they claimed the soils were globally correlateable, could enable some elucidation of the global climatic changes that occurred during that period.

The fact that they claim to have found 103 paleosols of both glacial and interglacial stages in warm Greece with its elephants and hippos, of all places, forces me to reject the latter claim as unsound. But, it must here be noted, I am simply rejecting Paepe et al.'s *interpretation* of the evidence; everything else seems perfectly sound.

The most interesting part of Paepe et al. is their effort to account for the origin of the red soils, loess, and aeolian dust in the first place. When they compared the Greek series with those of China's Central Loess Plateau and those of Africa in Burundi, Congo and Tanzania, they found them both similar, and time-stable (time-equivalent) which is not too surprising since they're all recent, obvious by virtue of the fact that they're lying on top of everything else. These red soils are also widespread in India, Australia, the Americas, West Asia (Middle East), and, as we know, all over the Sahara. In other words, they're a global phenomenon.

They are also found on the sea floor, and, again, all over the world. These authors thus consider that the dust, silt and loess must have had an extraterrestrial origin, and they are the first orthodox academics I've come across that are anywhere close to the truth of the matter, and do at least reflect what some of the ancient legends or "geomyths" had to say (which we've seen in rivers turning to blood, for example). They also, of course, reflect what Velikovsky the heretic had to say.

The global dispersal of red dust, more or less the same everywhere, forming soil or clay, also more or less the same everywhere, can only be properly explained by a generally acting agent that affected the entire globe during a brief event. It is also the only rational explanation for the (yellow, tan, brown) loess that is also widespread over the globe; it too must have an extra-terrestrial origin, as the idea that it blew down from the polar regions, as I noted earlier, does not make sense from a practical point of view, in light of the fact that everything is frozen up there and no wind blows very far southward and certainly not radially down every point of the compass.

The famous yellow loess of China gets its color from limonite, another iron mineral, and the tan and brown versions are simply mixtures, but the yellow and red are clearly related, and, as we saw at Pikermi, yellow silt overlies the red, with tan-colored also present. Apart from the lack of necessary winds, and, given that these silts are color-sorted to a great extent, the polar-source idea is obviously baseless in any case, as is any other wind source, considering wind, in general, is color-blind as far as I know.

Further, the standard explanation for silt, that of grinding by glaciers, is inapplicable in this case anyway, since polar icecaps do not behave like glaciers, and I would challenge any geologist to find a single such "silt factory" in either the north or south polar regions.

The explanation I offer is that the dust came from the tail of a comet, or other cosmic body, and showered down along with other material, and did so recently, since it lies on top everywhere.

This soil appears to me to be a massive dumping of cosmic dust at one go. The reason the red and yellow are generally separate but with some mixing as tan is likely due to some separation based on density or weight while travelling through space in the comet's tail. This suggestion comes to me originally from Ignatius Donnelly, discussed in some detail in his second book *The Destruction of Atlantis: Ragnarök, or the Age of Fire and Gravel* (1883).

While we won't get into it now, there are, of course, two hints in this direction that we are all familiar with, the red planet Mars, and the phenomenon of rivers "turning to blood." As we saw, in fig. 11-1, the Kokkinopilos area of Greece has been compared to the surface of Mars, and, as we also saw, there are many references to red rivers or rivers turning red, probably the most well-known being the "Nile turning to blood" in the Exodus story, as we briefly met in volume 1 with the aforementioned Velikovsky.

There are many others, and a related, and ongoing, phenomenon is the present-day existence of a large number of red rivers, whose water is red from the river flowing through areas rich in the very same red silt we've discussed in the foregoing. While rivers do, on occasion, turn red for various reasons, typically chemical or biological, the rivers referred to here are permanently red, from red silt or clay. And these red rivers are found all over the world also. Needless to say, we also know why China, with all its yellow loess, has a Yellow River.

Finally, we saw in Greece that red silt occurs in a number of places in the geological record, and from ages long before the recent present. The Pikermi fauna was buried in it, and it was a component of many sedimentary deposits, not to mention the red iron oxide giving its distinctive color to all that red sandstone to be found over huge areas elsewhere. It is also clear that water action was going on at the time this silt, with pebbles, was being deposited. Further, it must have been violent water action, in light of the fact that all those limestones were washed clean of any overburden prior to the placement of all that terra rossa.

Therefore, we can conclude that this earth has experienced massive influxes of red, yellow, and other silt, at various times throughout geological history, times that apparently saw major simultaneous water action. This would suggest that both Donnelly and Velikovsky were correct in their conclusions as to its origin, and that silt was likely a component of the dust "blown into both sides" during the *Titanomachy* and also the dust reported in the Buddhist book *Visuddhi-Magga*, as Velikovsky tells us. Hence, I must conclude that this red silt is likely one component of that "pestilence" that streams down from on high during that old Egyptian priest's catastrophes.

Chapter References

Boero, V. & Schwertmann, U., 1989, Iron oxide mineralogy of terra rossa and its genetic implications, *Geoderma*, Vol. 44, Iss. 4, pp 319–327.

Ignatius Donnelly, 1883, *The Destruction of Atlantis: Ragnarök, or the Age of Fire and Gravel*, D. Appleton & Company, New York; 2004, Dover Publications, Inc. Mineola, N.Y.

Fedoroff, N. & Courty, M., 2013, Revisiting the genesis of red Mediterranean soils, *Turkish Journal of Earth Sciences*, Vol. 22, pp 359–375.

May, Michael, 2006, The mystery of Terra, *The Western Scholar*, Spring 2006.

Merino, E. & Banerjee, A., 2008, Terra Rossa Genesis, Implications for Karst and Eolian Dust: A Geodynamic Thread, *The Journal of Geology*, Vol. 116, No. 1.

Muhs, D. R. et al., 2010, The role of African dust in the formation of the Quaternary on Mallorca, Spain and implications for the genesis of Red Mediterranean soils, *Quaternary Science Reviews*, Vol. 29, pp 19-20.

Olson, C. G. et al., 1980, The Terra Rossa Limestone Contact Phenomena in Karst, Southern Indiana, *Soil Science Society of America Journal*, Vol. 44, Iss. 5, pp 1075-1079.

Paepe R. et al., 2004, Quaternary Soil-Geological Stratigraphy In Greece, *Bull. Geol. Soc. Greece,* Vol. XXXVI, Proceedings of the 10th International Congress, Thessaloniki, April 2004.

Spratt, Capt. T. A. B., 1885, *Travels And Researches In Crete*, Vol. II, Appendix IV: On The Geology Of Crete, London, John Van Voorst, Paternoster Row.

Yassoglou, N. et al., 1997, The red soils, their origin, properties, use and management in Greece, *Catena*, Vol. 28, Iss. 3-4, pp 261-278.

CHAPTER TWELVE:

NEO-CATASTROPHISM

OUR EXAMINATION OF MUCH OF THE RECENT GEOL-
ogy of Greece, involving a fairly representative selection of books and
research papers on the various aspects and branches of it, has shown
that many geologists have had some trouble explaining the evidence
in strictly uniformitarian terms. We saw many references to various
and distinctly nonuniformitarian agents and processes drafted in as
replacements, i.e., flood events, alluviation events, meltwater events,
flash-floods, debris-flows, paroxysms, deluge and diluvial, etc., and
no hint of an everyday process anywhere.

   We saw that all these agents (we can't call them processes) had
to be repeatedly invoked to explain, or even just attempt to explain,
evidence difficult or impossible to explain by everyday processes. We
also saw much invoking of agents or processes not actually seen to be
in action in the present day, or any evidence that they ever existed, or
acted as described, either in the present day, or ever. We also saw a
major *lack* of evidence to indicate that any ordinary processes have
acted for any great length of time in the past or even in the recent
present, or have achieved much of anything, despite all the claims.

   Further, and again despite the implications of violence inherent
in the use of the terms mentioned, and the rather liberal use of them,

it is something of a surprise to me that all writers somehow managed to avoid using the words *catastrophe* or *disaster*, when these words are much more commonly used, and spring to mind much more readily, than words like *paroxysm* or *diluvial*. It's all the more surprising given that the vast majority of the papers and books consulted for this study were written since 1980, when the whole idea of catastrophes and catastrophic events was coming back into some respectability.

This, of course, was related to the Alvarez hypothesis of the dinosaur extinction, as we saw in volume 1 (Alvarez et al., 1980). And, as I also said at the time, this theory gave rise to some resurgence of catastrophist thinking, under the new term *neo-catastrophism*.

While the major focus of this new thinking was on the novel and morbidly exciting prospect of an asteroid impact as the cause of a major, and necessarily global, catastrophe that did away with the dinosaurs, as well as having various other effects, the door was also open to discussing other kinds of catastrophes, or "convulsive geologic events" as those involved preferred to phrase it. It seems academia still felt it to be wise (meaning, no doubt, career-wise) to avoid the dreaded (forbidden) term *catastrophe* with its global, biblical, and downright anti-uniformitarian connotations, and essentially "tone down" the catastrophic implications of neo-catastrophism.

However, it was quickly recognized by geologists that, while avoiding catastrophe on a global-scale, the concept of regional or local catastrophes, under another name, would be very useful for explaining many phenomena, without bringing the wrath and opprobrium of the powers-that-be down on the heads of such "disaffected" geologists. To that end, a segment of the geological community began to investigate phenomena that had been difficult to explain by adhering rigidly to uniformitarian methods and agents. Not that they recognized all that many phenomena as "hard to explain." Instead of slow and ongoing processes, the notion of a discrete "event" in the past was introduced.

Despite the fact that a natural catastrophe, or event, is a very easy concept to accept, in light of common experience with tsunamis, volcanic eruptions, hurricanes and landslides, etc., etc., all of which are discrete events, the "convulsive event" implied by the neo-catastrophists was a major break from uniformitarianism.

This is because all the events listed above are considered part and parcel of uniformitarianism, i.e., everyday events and any evidence for events of greater magnitude having happened in the past, would be (had to be) interpreted as the gradual accumulation of the effects of any number of such everyday familiar events, repeating over any particular, but always extensive, period of time, according to the standard doctrine. The difference therefore, really boils down to a matter of scale.

The Alvarez hypothesis is significant in that it was the first time in a long time that anyone had proposed the concept of a "global catastrophe." Of course, by "anyone," I here mean any respectable academic that's in much too powerful a position to be dismissed as the usual crank or crackpot. However, while Louis Alvarez was a physicist, his son Walter, a co-writer of the 1980 paper, was a geologist (Columbia University Website, 2016). And, while not labelled the usual cranks and crackpots, they were both considered so-called mavericks, and openly referred to as such.

However, despite their "maverickness" and the large amount of opposition, it was obvious that they did address a serious shortcoming in the science of geology. And, while many geologists were resentful at the intrusion of physicists and chemists into what they saw as their exclusive domain, it was obvious that something was badly needed to jolt geology out of its uniformitarian complacency, and that something would not be coming from the conservative, and perhaps resentful, doyens of the geology department.

However, as soon as Alvarez et al. loosed their cosmic dragon, anything was potentially possible, because there would be no getting that kind of useful agent, or any of its relatives, back into its cave, or bottling it up like a troublemaking genie.

Hence, other geologists in other fields began to re-examine various phenomena, and in the United States this led to the creation of a new division of the Geological Society of America devoted particularly to Sedimentary Geology. To get the ball rolling, an Inaugural Symposium was held in Orlando, Florida in (Oct.) 1985, the title of which was: "Sedimentologic Consequences of Convulsive Geologic Events."

A collection of papers, edited by H. E. Clifton (1988), was later published, comprising those presented at the symposium, plus additional contributions on other subjects. Most of the papers deal with what are simply scaled-up versions of the usual "events," such as volcanic eruptions, floods, tsunamis, submarine slumps and slides, and of course (theoretical) turbidity currents. The one paper (Pilkey, 1988) that most interests us with regard to the geology of Greece, concerns the filling of marine basins to generate what are termed "basin plains," which are simply basins with more or less perfectly flat surfaces extending over a very wide area.

This feature (or morphology) is characteristic of all the basins we looked at both in Greece and in the Aegean. The fact that the basins dealt with here are all submarine is not relevant to the actual fill mechanism that is required to produce the extensive flat, even surface, typical of these and similar basins on land. Essentially the same mechanism must be involved in producing the same result.

The paper opens with the following statement, in the abstract:

"The study of 13 modern basin plains ranging in size from the 200-km^2 Navidad Basin up to the giant (>100,000-km$^{2)}$ Hatteras and Sohm Abyssal Plains has revealed that these features owe their existence to large-volume turbidity currents capable of covering the entire basin floor. Such convulsive events flatten out the topographic irregularities formed by the deposition of small flows between the big events. Giant events maintain the flat plain floor" (Pilkey, 1988).

In Greece and the Aegean we dealt with relatively small basins but the Hatteras and Sohm are enormous.

The basins in question actually cover the entire ocean floor off eastern north America from Newfoundland all the way to Brazil and are more or less contiguous with minor ridges or islands between them. Some are large and open in the north, while others in the Caribbean are of limited extent and/or narrow, the most important of which being the Puerto Rico Trench.

We will examine these basins and trenches in much more detail when we study the Atlantic. For now, we are only interested in the physical appearance of the basins and the origin of the wide flat surfaces that are the subjects of this paper. Obviously, Pilkey ascribes them to turbidity currents.

Now, while I deny the reality of turbidity currents, they are, at the same time, in the minds of orthodox geologists, just large-scale sediment-laden floods that happen to occur under water. As we'll see, the mechanics are essentially the same as any sediment-heavy flow on land, like a very watery debris flow or lahar, for instance, and the whole *raison d'etre* of turbidity currents is to explain deposits in the sea, that if on land, could only be explained by a massive flood.

And, while denying massive flooding on land, it is as though by making up a new name and transferring the concept to the much more dense, and out of sight, medium of seawater, the physically impossible becomes magically possible. Invisibility, it seems, is very useful to the science of uniformitarian geology. To continue:

According to Pilkey, if a basin is fed by small flows in various places, then each flow will only cover a small portion of the basin surface and uneven deposition will result. The plain will no longer be a plain, and, obviously, multiple individual flows from different places into a basin cannot produce a flat planar surface. Pilkey says: "It is the occasional large flow that 'repairs' and flattens the depositional environment by preferential deposition in basin lows. Such large flows in most basin plains are truly convulsive events."

He concludes his interesting paper by stating that: "The mechanics of basin filling are such that a flat plain can only be formed if occasional events are large enough to cover the entire basin plain and fill in topographic lows produced by uneven rates of sedimentation of small events" (Pilkey, 1988).

He might as well be talking about the Greek and Aegean basins. When he says "small flows into a basin," he's here talking about small turbidity currents that he, and everyone else, believes occur between the big ones he's postulating here. Pilkey is, of course, assuming they produce the supposed unevenness that underlies his "basin-filling large flow."

If on land, and there are lakes in the basins, there are, of course, no flows into them; rivers drop everything at the edge, as we know, and nothing is transported into a lake in a basin. If it is the case of rivers flowing through a basin, then the topography will be uneven as is normal in all river basins and caused by floods that unevenly deposit material on the floodplain. However, he is, of course, and logically, correct that a massive flow completely filling a basin will produce a flattish floor over the entire basin.

In which case, repeating flat layers that cover an entire basin, which is what we saw in all the Greek basins, can only be produced by a series of depositional events large enough to cover the entire basin each time, i.e., large floods in this case. This is entirely what one can easily imagine as necessary to spread the material out in the even layers we found in those basins, while turbidity currents, which are nothing if not large submarine flows, and, though still theoretical, are held responsible for the basin-fill in the Aegean, as they are for the basin plains in the Atlantic.

Hugely massive floods, therefore, by inference and evidence, physics and mechanics, must have been what filled the basins of Greece as they swept through and over them, in the form of powerful sheet-floods of unknown depth and just as unknown lateral extent. However, they must have been massively deep and extensive.

We know that the basal layers in the basins we examined on Greece are composed of coarse clastic layers of conglomerates, gravels and sands, etc., all high-energy productions, and these are topped with finer silts and muds along with laterally extensive, and neat, sharply-defined layers of marls and lignites, features common to basins over a huge region.

The conclusion one is forced to arrive at is that Greece and the surrounding region was struck by a sudden and massive, catastrophic tectonic disturbance, that stretched and fragmented it. It was also struck more or less simultaneously by a number of enormous floods that severely eroded it and filled in the basins, and that uprooted vast amounts of trees, bushes, and plants.

The ultra-violence of the floods then tore the trees to shreds and macerated them, did much the same to vast numbers of savannah animals in the region, as well as an admixture of northern animals, and deposited everything in layers in the basins. The organic material turned to lignite, and interspersed with the lignite layers are layers of marl whose origins are unclear, even though orthodox geology presumes, as usual, that it is simply dissolved limestone from the local rocks, and denies pretty much all of the foregoing.

Chapter References

Alvarez, L. et al., 1980, Extra-terrestrial cause for the Cretaceous-Tertiary extinction, *Science*, Vol. 208, pp 1095-1108.

Clifton, H. E., 1988, Ed.: Sedimentologic Consequences of Convulsive Geologic Events, *Geol. Soc. Am*, Special Paper No. 229.

Columbia Website, 2016, October, Geologist Who Linked Cosmic Strike to Dinosaur's Extinction Takes Top Prize.

Pilkey, O. H., 1988, Basin Plains; Giant sedimentation events, *Geol. Soc. Am* Special Paper No. 229 pp 93-99.

CHAPTER THIRTEEN:

FLOODS OF GREECE

NOTWITHSTANDING THE GENERAL ORTHODOX denial of catastrophe, and in direct contradiction of their own uniformitarian paradigm, and all their theories on the geology of Greece, orthodox geology will, readily or reluctantly, admit that the region has suffered a number of major floods, some of which have names, as mentioned. In our study of Greece and the Aegean, we saw evidence that suggested that floods overwhelmed the region from a number of directions.

One from the northwest appears prominently represented in basins all over Greece, Attica, and the Peloponnese. This has been proposed by some geologists as due to meltwater "pulses," and by P. La Violette (2017) as being due to "mega meltwater waves" from the European ice sheet, the route of the water following the Danube drainage basin, presumably. It is, however, difficult to see how a flood from the ice sheet could fill the Danube basin so much as to overtop the Dinaric Alps and flood Greece, rather than simply flow on through the wide-open Danube valley and into the Black Sea. However, as we saw from evidence in northern Greece (Voidomatis valley, Pieria Mts.) some flood appears to have done so, but hardly from the Danube.

An unnamed flood from the south is indicated by those large, coarse, bouldery deposits in the mountains of Crete, thought by Nemec and Postma (1993) to be due to meltwater floods, due to a lack of alternatives, but most likely due to a large-scale marine inundation from the south. The evidence presented by Runnels et al. (2014) in the form of large alluvial fans with boulders, etc., leaves one in little doubt about the volumes and speeds of the waters involved, and not to mention Spratt's boulders.

It is, perhaps, possible that this flood may have been the returning ebb tide of the main flood from the northwest. Considering the direction, it is also possible that this was the flood that left sedimentary deposits forty feet above sea level in the Dardanelles, and containing Mediterranean molluscs, as reported by Suess (1904), but the molluscs could also have come from the west when the Aegean connection to the Black Sea was established, with a flood being implied.

Fig. 13-1: Giant ripples in the Chuya Basin, Altai Republic, Russia.
Image: Wikimedia Commons, CC-By-SA-4.0 Int., photo: Gaianauta.

A flood from the east is definite, as there is overwhelming evidence for a major flood east of the Black Sea. This flood is named the Altai Flood, and was supposedly due to a meltwater outburst from the Altai Mts. of Siberia. In any case, this flood is undeniable and its course can be traced through the Aral, the Caspian, the Manych Depression and on through the Black sea and out into the Aegean.

Huge scour channels and high terraces along with giant gravel ripples (fig. 13-1) are easily identified along its route. As we saw in previous chapters, terraces are simply abandoned floodplains elevated above the existing valley floor, and do not signify a previously filled valley, despite the orthodox "explanation." Here along the route of the Altai flood, there can be no question but that those high terraces were put in place by that flood. Further, there is a major body of sediment immediately outside the western mouth of the Dardanelles. This flood, of course, is thought to have derived from an ice-dam failure, and supposedly occurred at the end of the Ice Age (Baker et al., 1993; Rudoy, 2002) but we'll get to that.

Finally, a flood from the west is generally accepted by many geologists, though far from all, and there has been growing doubt about all, or parts of the theory, as well as the evidence. This flood is known as the Zanclean Flood, dated at 5.3 my ago. It is the flood that supposedly refilled the Mediterranean after it had closed and supposedly dried up to a desert.

The flood supposedly came over the Gibraltar sill and eroded it down (Hsü, 1983). It is considered to have occurred at the beginning of the Pliocene. The main reason the Mediterranean is thought to have dried up is that thick beds of salt and shallow-water evaporites like gypsum were found in what are now deep basins. In light of the subsidence of Aegeis, the fact that the basins are deep at present does not necessarily mean they always were.

However, desiccation is not necessary for the deposition of salt beds. Salt precipitation will occur where a persistent but restricted inflow of ordinary seawater but little or no outflow and ongoing evaporation all exist. This leads to the exsolution of salt and its accumulation on the sea floor (Garcia-Castellanos et al., 2020). And this is the case in the Mediterranean, which has a restricted inflow in the west and no deep outflow due to the shallow depth of the sill at the straits of Gibraltar, and no other outlet.

Almost needless to say, most of those who accept the theory, consider this flood to have progressed relatively slowly, though still "catastrophically" by normal geological standards, by erosion of the sill between the Atlantic and the Mediterranean, even though no erosion can go on at depths below about 30 m (100 ft), if even that, because there is no wave effect and nothing else operating at this or greater depths. This suggestion is simply the same as, and likely the inspiration for, Ryan & Pitman's imaginary flood over the sill of the Bosporus, which, despite the acquaintance of western geologists with the Altai Flood evidence, some still entertain this idea.

The desiccation theory suggests that the Mediterranean dried up because Africa pushed up against Europe (as we saw) and closed off the Mediterranean in the west and east, presumably during the formation of the Alps. Millions of years later, further movement involving faulting, collapse and erosion, or some such mechanism, poorly understood or vaguely guessed at, caused the Gibraltar sill to breach and the water of the Atlantic to flow into the Mediterranean.

This is the usual description of the event, but Abril & Periáñez (2016) helpfully add that the Mediterranean Sea, in effect, "captures" the Atlantic Ocean (in a most audacious act of "piracy," I must say) and this is another one of the most absurd statements I've yet met.

Of course, if some event, currently beyond our knowledge, were to seriously disturb the Atlantic Ocean, then it is not hard to imagine that ocean sweeping masses of debris in over the sill, eroding it down and filling up the Mediterranean very quickly. As a matter of interest, Garcia-Castellanos et al. (2020) tell us that Pliny the Elder (~77 AD) in his *Historia Naturalis* reports on a legend from Iberia, telling how the Mediterranean was born.

According to the legend, the Mediterranean was deserted and cut off from the Atlantic, until Heracles dug an inlet with his sword between Jebel-el-Mina in North Africa and the rock of Gibraltar. This allowed the Atlantic Ocean to flow into the Mediterranean Basin, where "it was before excluded," thus "changing the face of nature" (Pliny).

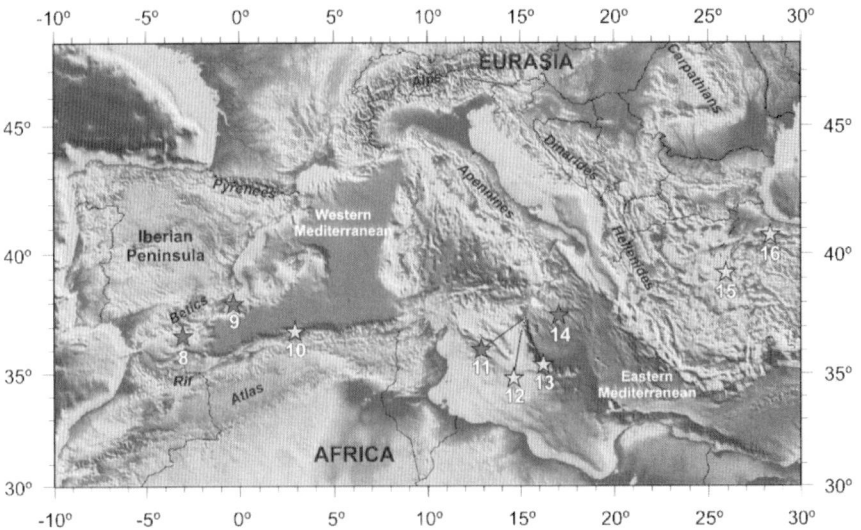

Fig. 13-2: Topographic map of the Mediterranean. Bathymetric data from the European Marine Observation and Data Network. Image: A. Fontana. (Benjamin et al., 2017).

Taking account of the fact that the 5.3 my age for the Zanclean Flood comes originally from Charles Lyell's imagination, and uniformitarian theoretical requirements, Pliny's legend might be more correct in suggesting a date for the event considerably closer to today, and it does at least give one possible and good reason why the straits of Gibraltar have long been known to the people of the Mediterranean as the "Pillars of Heracles." See Pliny's *Historia Naturalis* and needless to say, we recall Hesiod's *Theogony* here. Pliny's legend, of course, may simply be an etiological myth describing the reopening of the straits.

The submarine topography of the Mediterranean is such that the sea is divided into two major basins, a western one and an eastern, separated in the area of Tunisia-Sicily-Italy, where there is a wide, shallow ridge or plateau, as shown in the 3-D map of fig. 13-2.

As I said above, desiccation is not necessary for the production of beds of salt, and it is somewhat difficult to understand how the entire Mediterranean could possibly dry up, given that it is not in a desert region such as is the Sahara, and half the rivers of Europe flow into it, as well as the Nile and many rivers of Russia, depending, of course, on a connection between the Black Sea and the Aegean region, either before or after the sinking of Aegeis, and Suess (1904) mentioned a river from the Black Sea flowing over Aegeis to the Mediterranean.

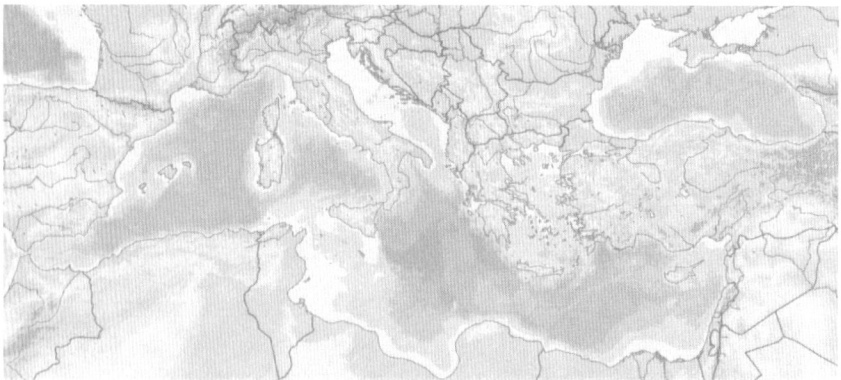

Fig. 13-3: Map of the Mediterranean Region showing major and many minor rivers and the shallow region between Tunisia/Libya and Sicily. Image: Wikimedia Commons, CC-BY-SA-3.0 Lic., photo: Nzeemin.

The northern shore of the Mediterranean has, from west to east, a series of rivers flowing from the north, as can be seen in fig. 13-3, and many of these rivers are of large size, e.g., Ebro, Rhone, Po, along with any number of smaller ones, as well as the waters of the large European, Ukrainian, and Russian rivers that come out through the Black Sea, i.e., the Danube, Dniester, Southern Bug, Dnieper, Donets, and Don, as well as the Rioni in Georgia. On the southern shore we have the Nile, the largest, or second largest, river in the world. It is thus difficult, given present conditions, to envisage how the Mediterranean could dry up.

Even if those above-named rivers didn't exist in the long ago, other rivers would have, because it has always rained and snowed in northern latitudes, and the water will always find its lowest level. Since, as we can see from the map, the Mediterranean itself is the lowest region of Europe/Russia/Anatolia/North Africa, it is safe to assume that it would always be a collecting basin for the rainwater runoff and meltwater of vast areas of continents, including many mountain ranges.

It is highly questionable, therefore, as to whether the "Messinian Salinity Crisis" was ever a reality, or whether it is just another figment of the geological imagination, as, in fact, many geologists consider it to be.

The first "evidence" of the "Messinian Salinity Crisis" comes from the work of the Deep-Sea Drilling Project in the Mediterranean in 1970 and 1975, under Kenneth Hsü, of the Swiss Federal Institute of Technology in Zurich, assisted by Bill Ryan of Ryan & Pittman Black Sea Noah's Flood (1998) fame. It was, of course, drilling by the *Glomar Challenger* that revealed all the so-called evaporites.

While the drill core did discover various types of chemical precipitate, and supposedly sabka-like deposits, the conclusion of a former desert is based on the finding of "well-rounded gravels" and "massive red and green silt beds" (Hsü, 1983, p. 148) near the Balearics, but far from the coast, at which point Hsü rightly wonders: "What are they doing here?" (A "sabka" is a tidal flat, estuary or lagoon in an arid environment where evaporites develop.)

Red and green silts suggest oxidation, and, as far as everyone at the time was concerned, typical of a desert. Were the Mediterranean located 20 degrees further to the south, where the Sahara Desert is now, and much shallower also, it would, perhaps, be possible for it to dry up, but it would likely also require that much less water was flowing into it, which would seem reasonable. It is also, I consider, an absolute necessity that the Mediterranean was a lot shallower, as I'll explain shortly.

We discussed the red silt at length earlier, and its relation or otherwise to a desert. As we've seen, it's only a matter of traditional first-glance uniformitarian assumption. Given that the gravels imply rapid water movement far from the coast, I'd have to consider it a bit early to be drawing such conclusions from the limited amount of evidence so far obtained. However, Hsü was writing in 1983, so the red color was still conventionally associated with deserts.

The actual evaporites found were gypsum and anhydrite and these do form in shallow water as a result of evaporation, and they are usually found near the coast, in bays and lagoons or estuaries, etc. We'll discuss them further later when we get a better idea of exactly what was found in the deep basins of the Mediterranean.

I would also not be in any way certain that the floor of the Mediterranean has always been stable, and it may have risen and fallen at times, since we have seen much evidence of recent subsidence in the region. It is, at the same time, logical to conclude that the sea floor would have been raised up by the compression caused as Africa pushed against Europe, which clearly has occurred, and it more than likely would have subsided again when the stretching and subsidence of the Aegean occurred, i.e., as Africa pulled away again.

One of the main problems with Hsü's contention is that were the Mediterranean to dry up, based on its current water volume, a layer of evaporites averaging between only 20 m and 50 m thick would be deposited over the floor, depending on the depth; the deeper the basin, the thicker the deposit. However, the thicknesses of the evaporite beds in the different basins reach more than 1,500 m (Gargani et al., 2008).

Various hypotheses have been proposed to get geology out of this conundrum and maintain the theory. As is often the case, and no surprise, Hsü's large singular event is instead transformed into a series of smaller events stretched out over time, such that the new scenario is one where the Atlantic sill is repeatedly breached and closed again, and another, proposed by Gargani et al. (2008), whereby a breach is repeatedly made and closed again with the Red Sea at Suez (implying that Africa and Europe have been bouncing off each other?). This, obviously, is all simply pure invention.

The difficult issue with regard to the complete evaporation of the Mediterranean is that, while evaporation currently exceeds input, and the difference is made up by Atlantic water flowing in, as the sea dries up, its level naturally falls, as does its areal extent. The sea, therefore, becomes progressively smaller and thus the area of sea surface available for evaporation also decreases in parallel.

As shown in fig. 13-4, at 1500 m, the Mediterranean is divided into two distinct basins with a wide ridge or plateau between Africa and Italy, and incorporating Sicily and the Maltese islands. While the area of the water's surface becomes smaller, there would be no reduction in the amount of inflow from rivers, and thus to accomplish almost complete desiccation, evaporation would theoretically have to increase accordingly, which seems impossible without massive amounts of extra heat, for a start.

Fig. 13-4: Digital elevation model of the Mediterranean during the Messinian Salinity Crisis. The blue areas define the 1,500 m depth contour which is considered the extent of Messinian sea level and signifies a pull-back from the idea of total desiccation. Image: Loget et al. 2006.

There is one other major issue with regard to the desiccation of the Mediterranean that has not been mentioned by anyone, as far as I know. That is that as the sea surface falls below sea level and further down, air pressure increases and a situation such as prevails today at the Dead Sea will develop in the Mediterranean before very long. The Dead Sea and its surroundings are covered with a haze from evaporation, and, because this haze-filled air is so dense and there's no wind, it doesn't disperse.

This thick haze hangs in the air and prevents the sunlight reaching the surface, much reducing the temperature and the sun appears only as a glow in the sky. Therefore, evaporation will necessarily reduce as the water level falls. The Dead Sea is 430 m below sea level, and, because of the haze reducing the sun's intensity, one doesn't get sunburn, and I'm pale as a ghost and was there in July when the temperature elsewhere in the area was over 40° C. Hence, if the sea level were to fall to the 1,500 m of fig. 13-4, evaporation would reduce a good deal further or more likely stop well short of that.

This is a physical principle we're all familiar with. The effect is similar to the principle of operation of a pressure cooker or steam boiler. As one increases the pressure, more and more energy is needed to boil or evaporate water. The steam is now at a high pressure and in the case of a steam engine, the high-pressure steam can be used to do work, i.e., drive a locomotive. In the case of the pressure cooker, the water boils at a higher temperature and hence cooks the food more quickly.

It is the opposite of what happens on climbing a high mountain or flying. The reduced air pressure allows water to boil at a temperature well below 100° C. It is also a principle utilized in industry, particularly food processing. Fluids such as juices or process-water are concentrated by heating in vessels under partial vacuum. The water boils at a temperature of only about 40° C and hence much less energy is required.

The case at hand is characteristic of the steam boiler, or steam engine. Given the steadily increasing air pressure, much more energy will be needed as the water level falls further and further below sea level. Hence at any given temperature, therefore, much less water will be evaporated at the excessively low levels postulated for the desiccated Mediterranean proposed by this theory, and I would consider that a limit would quickly be reached. Furthermore, evaporation becomes more and more difficult in the increasingly saturated atmosphere that would be created.

With the increasing air pressure and the heavy haze created as the surface falls, it will also be more and more difficult for that heat just to reach the surface to actually do any evaporating. Hence, I would have to suggest that the only way to desiccate the Mediterranean is for it to be a very shallow sea. And if it was a very shallow sea, the evaporites found by Hsü are easily explained. Further, the deep-water molluscs also found by Hsü were likely brought in by the refilling Atlantic, or were there before the Mediterranean was closed off, as Hsü himself suggested.

When discussing these fossil molluscs, Hsü refers to them as either Pliocene or the earlier Miocene, which he says largely died out during the Pliocene. His "Messinian Salinity Crisis," therefore marks the boundary between the earlier Miocene and the later Pliocene. He then declares that: "This finally provides the rational basis for Lyell's distinction of the more (*Plio-*) from the less (*Mio-*) recent faunas." One must immediately wonder what "irrational" basis Lyell used instead, especially since he was busy defining his Tertiary Period at the time, and ass-slapping long ages onto everything in sight, ages that Hsü is here suggesting might be irrational, and which I have long thought were simply suspicious.

Also, we know what the balance between inflow and evaporation for the Mediterranean is today, and it would have to be known for the Miocene also before any claim at all could be made for desiccation or otherwise. And this balance has everything to do with the heat energy received by the Mediterranean, a question for physics, but of no use unless the Miocene climate and the size of the Miocene Mediterranean are known with some certainty, and not just theorized, something which is not really much considered. Logically enough, depending on the equation, there comes a point where inflow exceeds evaporation.

We saw that the Pliocene animal fossils of Greece clearly implied a savannah environment and it was, therefore, much warmer in Greece and the general region in the Pliocene, though the climate of the Miocene is uncertain. If it was like today, rains would have been much the same as they are now. Otherwise, were it sub-tropical, then monsoon rains could be expected.

There being no evidence of a cooler climate, therefore, the Mediterranean would have received plenty of water. Given that basins existed at the time, with saltwater in them, and I can easily accept a shallower Mediterranean, I cannot, however, see a good argument for complete desiccation, based on the known physics of evaporation versus pressure, unless the Mediterranean were no more than a few hundred meters deep.

This means that, at present at least, I would consider the complete desiccation hypothesis to have little basis in evidence, theory or physics. At the same time, the age of 5.3 million years I would take with a large chunk of the salt Hsü found with the coring tube, without a shred of evidence to support it, given that everything we saw on Greece is close to the surface, suggesting it's not very old, and nowhere near millions of years, despite what orthodoxy says.

Mind you, Hsü's 1983 book, *The Mediterranean Was a Desert* is quite a good report on the drilling program carried out, and, in fact, in the book, William Ryan expresses his opinion that the Mediterranean, or its basins at least, were formerly shallower (Hsü, 1983). I would have no trouble agreeing with Ryan on this point, considering all the evidence we've seen for stretching and subsidence in the Greek and Aegean region. Since the Mediterranean is right alongside, it could hardly have escaped. However, Hsü rejected Ryan's suggestion as he could not envisage how the Mediterranean could have been shallower.

It is clear, of course, from where Ryan got his inspiration for his and Pitman's Black Sea flood hypothesis. The foregoing, however, is not to say that no flood occurred in the Mediterranean. As I've made clear by now, I consider that a flood did occur; it was just an altogether different flood, and at a much more recent time, and there's no shortage of evidence on Greece that a flood came in from a westerly direction.

I mentioned at the beginning of this description of the Zanclean Flood that there is both significant and growing opposition to the theory that the Mediterranean dried up more or less completely, which would have necessitated a water-level drop of 2 to 3 km (6,000 to 9,000 ft.) which, as I've said, I would consider essentially impossible. The 50-year history of the theory, its development, its problems, and its current status, have been well-described by G. Battista Vai (2016).

In his paper, he tells us that, apart from a few diehards like Ryan (2009), etc., most geologists have given up on the idea, due to all the evidence that has been found since Hsü first proposed his theory. It turns out that, following decades of extensive investigations, much of the evidence marshalled in support of the theory, not to mention underlying assumptions, have been found invalid, disproved, or baseless.

According to Vai (2016), the general consensus nowadays is that the Mediterranean was not cut off completely from the Atlantic Ocean (I beg to differ) and did not dry up. Instead, the ". . . scenario has passed from desiccation to normal marine inundations, ingressions, influxes in a mainly hypersaline (salt-rich) water body. These are the new keywords 'overflooding' (a little joke on Vai's part) more and more what was displayed as a desert" (Vai, 2016).

Here, of course, a "normal marine inundation" or "ingression" or "influx" means the usual slow sea level rise. The need for a "hypersaline" water body is purely as a source for the massive salt deposits, which are difficult to explain by either scenario.

Vai goes on to cite a number of workers who have claimed that the actual drop in level in the Mediterranean was either a few hundred meters (Clauzon et al., 1996; Roveri et al., 2001) or absent altogether (Krijsman et al., 1999b; Roveri et al., 2014b) and therefore, these workers claim that the gypsum units were deposited in a deep-water basin. In which case, as Vai puts it: ". . . the charming desiccation theory is turning into a tale."

Evaporites in a deep-water basin, however, is clearly contradictory, water doesn't evaporate from the *bottom* of the ocean, as we all know well. Ryan, to my mind, was perfectly correct; one can only get evaporites in shallow evaporating seas. Shallowness, to produce all the salt deposits by evaporation, was the very reason Hsü proposed the desiccated Mediterranean idea in the first place.

However, it was found that many so-called sabka deposits in coastal canyons of the Levant turned out to be clastic sulphates (broken-up gypsum/anhydrite) resulting from "sub-aqueous mass-wasting," which means underwater landslides, slumps, or gravity-flows, as were other gypsum deposits in other basins (Vai, 2016). Roveri et al. (2014), citing Hsü (1983), report that no turbidites were found in the deep basins, so at least we're going to be spared the turbidity-current hypothesis, for a welcome change.

However, one question here is where did the "clastic sulphates" come from? They are now found in deep water, and must derive from shallow water. Hence, the evidence suggests that they were produced during some kind of disturbance, and this disturbance was likely related to a subsidence of the Mediterranean sea floor, which subsidence logically brought the clastic sulphates down with it.

The Roveri et al. (2014) paper is an attempt to map out the future of research into the Messinian Salinity Crisis after, as they say: ". . . a long period of debate over several entrenched but largely untested hypotheses . . ." In other words, for the past 50 years geologists have been debating this theory without, apparently, checking the validity of the hypotheses on which it is based.

In his paper, Vai (2016) tells us that the general Messinian Salinity Crisis theory, as first proposed, and the Zanclean Flood hypotheses, are both now more or less defunct. The main reason given by a number of writers for the lack of clarity, or progress, over the last 50 years is the fact that the question was addressed by two different scientific communities, one on land and the other oceanographic, who, as usual, do not, or did not, communicate with each other.

Thus, it has been difficult to establish a unified scenario (consensus of opinion) such that: ". . . Messinian research has been characterized by endless controversies among different research groups . . ." (Roveri et al., 2014). And these writers call for the Messinian Salinity Crisis question to be considered an excellent opportunity for collaborative work between the two separate scientific communities, to achieve a full understanding of the event (Roveri et al., 2014). One really must wonder why they weren't already collaborating.

One such controversy is pointed out by Vai (2016) with regard to the climate of the times. Some authors say the MSC was related to a cool or cold climate whereas others say warm and subtropical, even up as far as northern Italy. Obviously "cool or cold" is hardly the type of climate calculated to induce a massive amount of evaporation, and we've already seen in the basins of Greece that a savannah environment existed, teeming as it was, with subtropical animals. We saw no evidence in Greece or met any suggestions that the earlier Miocene was any different.

Nothing we saw in the Greece of the Pliocene suggested anything other than a warm subtropical climate. Hence, the evidence, and the logic of physics and astrophysics, favors the warm climate option. Another major problem with the desiccation theory is that the continental geology of the lands surrounding the Mediterranean does not, apparently, show much evidence of significant sea level fall, and quite the opposite in many places (Saner, 1928).

While few now accept the desiccation theory, there is, at the same time, plenty of evidence of shallow-water evaporation around the Mediterranean coastline, in the form of anhydrite and gypsum deposits. What these also show, however, is that the Mediterranean turned from a saltwater sea to a brackish or freshwater lake.

What identifies them as having formed in fresh or brackish water is their fossil content, all fresh/brackish forms, though since they are found all around the Mediterranean (Grossi et al., 2008) and brackish water is typically only found in the mixing zone in estuaries or enclosed bays, some writers have considered that the entire Mediterranean became brackish and that the Mediterranean was entirely closed off from the Atlantic (again, as Pliny the Elder and Hesiod said). The freshening event is referred to as the "Lago-Mare" event, and it is undeniable, but does add another element of complication to the sequence of events.

Grossi et al. (2008) claim that a period of freshening and deepening was followed by a rapid increase in both salinity and depth, while Caruso et al. (2020) mention the alternative of a high water level in an almost fresh Mediterranean, and the maintenance of a constant connection to both the Atlantic Ocean and also to the Paratethys in the east. The Paratethys is, of course, the area of lakes referred to by Spratt and Suess.

The Paratethys to the east is here considered the source of the fresh water, via ". . . massive inputs of Paratethyan waters rich in organisms typical of fresh and brackish water." They also mention very unstable "lacustrine" (freshwater) to brackish environments with various deposits exhibiting sub-aerial exposure. The term "massive inputs" sounds very much like floods from the east, like, for example, the Altai floods.

While telling us that the Messinian/Zanclean contact zone is well-marked by a "clear erosional surface that grades laterally into megabreccia with reworked gypsum blocks," they also state that the Messinian-Zanclean boundary corresponds to a major change. Marine re-flooding occurred "abruptly" and there was an "instantaneous" restoration of marine conditions at the onset of the Pliocene (Caruso et al., 2020). This sounds very much like a massive flood to me, or a repeating flood, since the writers also mention "five lithological cycles," cycle, of course, being the operative word.

The big question here is as to how the Mediterranean Sea can change from saltwater to freshwater and back again, seemingly very rapidly, as the evidence, and these writers, are suggesting. From the "massive dumps" from the Paratethys, we have Vai suggesting a source through Slovenia, but he also suggests that the Paratethys basins were elevated higher than the Mediterranean and dumped all their water into it too, to produce the wide and rapid invasion of freshwater (a huge flood?) seen during this "Lago-Mare" event.

He also suggests that the Adriatic may have been an entry point for cold water draining off the Alps and sinking to form deep cold water at the bottom of the sea. Unfortunately, this is a physical impossibility, as freshwater will not sink through saltwater, due to the much higher density of saltwater. Freshwater will always "float" on saltwater with only a minor amount of mixing, if there is some turbulence. The Amazon for example sends fresh water two hundred miles out from land before mixing with the ocean water.

In fact, Grossi et al. (2008) discuss the degree of salinity of the water, and describe it as variable, spanning from "oligohaline" to "mesohaline," which means from an almost negligible to a low salt content. Hence, the Mediterranean appears to have been closed off from both the Atlantic and Indian Oceans and seems to have consisted of a layer of fresh surface water of some unknown depth, below which was saltwater.

This would go a long way to explaining the presence of freshwater fossils at the surface and on land around the coastlines, and marine fossils, fish, molluscs, etc., in other places, such as those reported from Italy by Carnevale et al. (2018) and as we'll come across again elsewhere. It would also explain the marls found off the Patras Gulf in Chapter 6.

The only reason left for adhering to the desiccation theory, with its extreme low sea level, is the presence of all the submarine and buried canyons around its perimeter. As we've seen, orthodox geology considers that the only way to generate a river valley is for the river flowing in it to carve it out. The low sea level is therefore necessary to provide a means for the water to flow downslope and thereby erode the canyon. The most famous are those of the Nile and the Rhone, but there are approximately 27 around the Mediterranean. Many, or most, are already filled up with sediments, and a typical canyon can extend hundreds of miles inland. To me, they look like buried fjords, but we'll leave this issue for later.

Personally, however, I would have no problem accepting that at some time in the not-too distant past, the Mediterranean was isolated and may even have comprised two separate, and shallower, basins, a western one and an eastern one, similar to those shown in fig. 13-4. We will disregard the depths for the time being, as we have no accurate idea of how much uplift and/or subsidence any part of the sea or the surrounding land has undergone in the recent or semi-recent past. We just know that there have been land movements. As we can see (fig. 13-4) the entire Aegean is dry land, as it used to be up to 12,000 years ago.

Apart from that, the area around Sardinia-Corsica-Italy is also very shallow, as is, very significantly, Tunisia-Libya-Malta-Sicily-Italy, the latter which would have formed a clear and distinct, wide land bridge to Africa, the better to enable all those savannah animals to cross back and forth. It may also, therefore, be a good place to look for signs of human habitation, on the sea floor that is, and, as we remember from the reports of classical Greeks in volume 1, Sicily was attached to Italy until recently.

The only problem with the land bridge idea is that the Sahara Desert is south of it, and, because it's bone-dry, the Sahara is almost as much of a barrier for animals to cross as is the Mediterranean. Conditions, would, therefore, have to have been different in the Sahara region also, which they were, in fact, but not so long ago as the Pliocene, as we'll see. In volume 4, we will see that the Sahara was a very different place not too long ago.

One other flood, mentioned above, that has been the subject of a lot of recent debate is that which is variously referred to as the Black Sea Flood, (the real) Noah's Flood, the Black Sea Deluge, etc. This theory, proposed by Ryan et al. (1997) posits that, as sea level rose after the end to the Ice Age, the Mediterranean overtopped the sill between the Aegean and the Black Sea at the Bosporous.

The Black Sea was, according to these writers, much lower at that time, and the slowly rising surface waters of the Mediterranean leaked over the sill and wore down a canyon, in solid rock, on the inside, and "rapidly" (i.e., at a hypothetical rate of mere 10 to 20 cms [3–6 ins] per day) filled the Black Sea depression. From everything we've discussed in the foregoing pages, we know that this water, leaking in slowly, could not have been carrying abrasive debris of any kind, and, quite apart from that, few geologists accepted this hypothesis, for any number of reasons.

Practically everyone else who had studied the Black Sea and the Bosporus, especially the Russian and Turkish geologists who had spent years at it, considers that the water went in the other direction, which the evidence-supported, and undeniable, Altai Floods would clearly suggest. This is also in line with Spratt (1865) and Suess (1904) and many others who agreed that the Black Sea was then a freshwater lake fed by the rivers of Russia that drained out through the Bosporous and, via one or more river(s) flowing over the then subaerial Aegean Platform, reached the actual Mediterranean far to the south in the region of Crete or Rhodes.

The suggestion that it was Noah's Flood was made by Ryan and Pitman, a co-writer of the original paper of 1997, in a book of that title they both published in 1998. We all know the story of Noah's Flood, and it is a bit much to suggest that the Mediterranean, leaking into the Black Sea depression, could in any way, shape, or form be confused with that story. It seems to me to be another example of "alternative theories."

The method used here is one I've mentioned a few times and is the same as the one used vis-à-vis Atlantis and other similarly awkward issues, and that is of burial under a mass of drummed up "alternatives" to ensure a never-ending debate, leading to no serious investigation and no answer at all.

In terms of the evidence we saw on and around Greece for three or four floods, it is clear that two floods stand out as much larger and more powerful, and as having had much greater effects than the others. These are, firstly, the flood from the northwest, much the more powerful, widespread and damaging over much of Greece and Aegeis, leaving behind enormous amounts of sedimentary deposits of all kinds and coarseness over very wide areas.

As mentioned, basins over a great area of Europe, Russia, and Türkiye all appear to have been affected in much the same way, and other areas further to the east may also have been.

Occam's Razor dictates that the condition of all the basins should be due to the same cause. Subsidence of the Aegean seems also to have been associated with this flood, in that the basins of the Aegean all appear to have been filled from the north with coarse terrigenous clastic material along with terrigenous organic material forming sapropel layers. We recall that the rock layers in the Aegean were continuous prior to the breakup and subsidence that formed the individual basins, and the infilling appears to have occurred immediately thereafter.

This, evidently, was the earliest of these last four floods, as the material deposited by this flood in the basins lies directly upon bedrock. This bottom layer is always substantial and comprised of coarse, cobbly conglomerate, as we saw, very commonly found lying directly on eroded bedrock the world over.

It is called a "basal conglomerate," as mentioned many times in the foregoing, and may be loose or consolidated. It is always lying on an unconformity, i.e., an erosion surface forming the bottom or "base" surface for many, if not most, coarse, clastic, sedimentary formations, whether loose or consolidated.

When we discussed "peneplains" earlier, I did not mention that, although it is true that no peneplains are found in the process of formation today, features taking the form of these eroded surfaces, exactly as peneplains are described, are to be found throughout the rock record, always comprising unconformities. Yet again, and contrary to the "principles of geology," they are simply never found to "be in action at the present day," so to speak.

Academic geology does not have an explanation for the "basal conglomerate" because it is always found lying on an erosion surface, which, almost by definition, means a bedrock surface; it is always coarse and cobbly, and may even have large boulders included; it is often of great thickness and areal extent, and obviously bespeaks water action on a scale orthodox geology refuses to contemplate.

From all of this and what we've seen earlier, we can conclude that the agent that transported all the conglomeratic material in Greece, the Aegean, and everywhere else, was the same agent that cleaned out and eroded the bedrock floors of all the basins, prior to deposition of that conglomerate. The flood obviously used the transported clasts to do the actual eroding, prior to the floodwaters slowing down enough to begin depositing the conglomerate and progressively finer material as the flood slowed progressively down.

This is the (actually others' and my) correct explanation for the peneplains and basal conglomerates found throughout the rock record; the one short-term operation is the simple solution; one agent, a major flood, and that agent only, acting massively, widely and rapidly, formed the peneplains by erosion using the material it subsequently dumped on those peneplains to form the basal conglomerates.

This, by the way, also explains the "alluvial" coverings on the peneplains and strath terraces that are such a long-time puzzle, but only from a uniformitarian point of view. It is also obvious that the sediment layers in the basins are due to the back-and-forth, or repeated, motion of floodwaters, heavily charged with calcium carbonate (marl) and with vast amounts of organic debris floating on top. As the ebbing and flowing of the flood subsided, the layers were deposited one atop the other, and all in neat, extensive layers, as we've seen, as logic dictates, and as at least a few Greek geologists agreed.

This flood I consider to be the oldest for which evidence is to be found in Greece, i.e., the conglomerates that lie directly on eroded bedrock, implying that this is the "basal" flood that swept that bedrock clean of any existing material, and further eroded it. As to when it occurred, I consider that it was possibly ". . . the greatest deluge of all . . ." and coincided with the end of Atlantis and the Ice Age.

It is the one responsible for much of the damage done to Greece, including the erosion of the mountains, the cutting out or enlarging of the river valleys, which were most likely roughed out first by faulting or tectonic disturbance, and the first infilling of all the basins, typically with three stacked clastic units, signifying three initial ebbs-and-flows of coarse material followed by many more to deposit the marls and lignites.

The flood from the west is more of a problem to explain. Its signature was most distinctly found in the basins of the Peloponnese, such as the Megalopolis, Athens and Mesogea as well as in some other areas, such as the Mygdonia Basin in the north. As we saw, the evidence for this flood lies directly on the evidence for the earlier flood, giving a very large-scale version of a typical stacked flood sequence.

As we saw at Pikermi in the Mesogea Basin, the overlying Pleistocene was easily field-confused with the underlying Pliocene, supposedly millions of years apart in time, but no deposits of any kind were reported from in-between, to represent all that time. Further, there is the issue of the similarities in animal and plant fossils, though one is supposedly of the savannah and the other of the Ice Age, yet they're almost identical, but with a slight admixture of somewhat more northern forms in the upper formation, which is considered Pleistocene. This is a very apparent contradiction, which is difficult to account for, other than by mixing by a well-traveled and widespread flood.

I am inclined to conclude that what we are dealing with here is either two massive floods separated by a period of time, where the first occurred at some indeterminate date, approximating 11- or 12,000 years ago, and the second one, at about 5,000 years ago. The second eroded the top of the first to eliminate some, or all, of the intervening deposits. That, or the evidence represents two waves of the same flood. The initial flood was from the north-northwest, while the wave from the west was a somewhat later and somewhat less powerful expression of the same massive disturbance (likely) in the Atlantic, attendant perhaps on the sinking of an island continent about 11,500 years ago.

At this point, however, it is far too early to identify anything securely, so the foregoing is only a somewhat more evidence-based speculation on my part. The fact is that despite the arguments about an empty or full Mediterranean, the evidence for a major west-to-east flood across the entire sea is unequivocal. That it has been called the Zanclean Flood and dated to 5.3 my ago is irrelevant to me. The fact remains that a flood occurred, as shown in fig. 13-5.

The series of diagrams and images included here collectively show evidence of major water movement from west to east across the Mediterranean basin. Fig. 13-5 shows the channel over the sill at the Pillars of Heracles, and the spread of the flood in the Alboran Basin immediately inside. While the presumption is that the water entering the Mediterranean eroded the channel over the sill, it is well to note that the Africa/Europe contact zone is a highly active tectonic area and lies at the eastern end of the Azores-Gibraltar Fracture Zone (also known as the Gloria Fault Zone) see for example fig. 13-7.

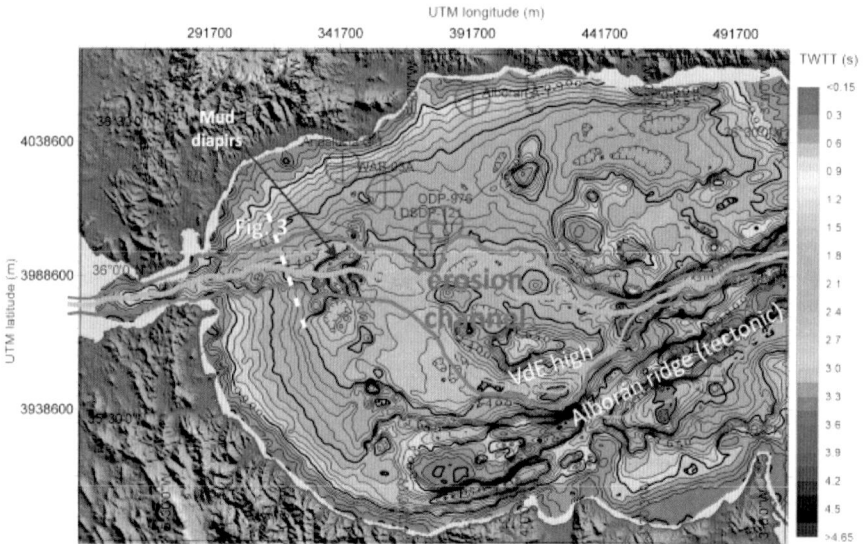

Fig. 13-5: Map of the Pillars of Heracles Area showing erosion surface and flood channel due to the Zanclean Flood, and the main flood-path in orange. Image: Garcia-Castellanos et al. (2020).

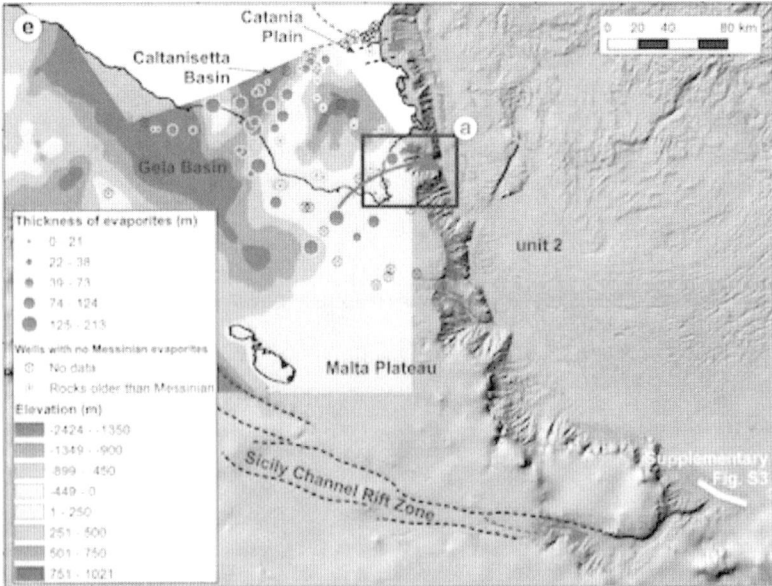

Fig. 13-6: Grey-shaded map of the Malta-Sicily Plateau showing topography, flood route and flood deposit. Image: Micallef et al., 2018.

This would suggest that the basement, which is the sill beneath the Straits of Gibraltar is almost certainly broken up by faulting, and the flood channel is not simply a function of erosion by leaking seawater, which, as I said earlier, it couldn't possibly be with the sill at a depth of 300 m or 900 ft., well-below any possible wave or other water effects.

It is instead I would think, like many similar phenomena, a case of a faulted channel, with subsidence, which was subsequently widened and deepened by a massive flood pouring through the faulted breach, while carrying any amount of debris, etc., swept up from the Atlantic side of the Straits, and therewith eroded the rock to give the Straits their current submarine topography, and, in the process, likely deposited all those gravels that Hsü found near the Balearics.

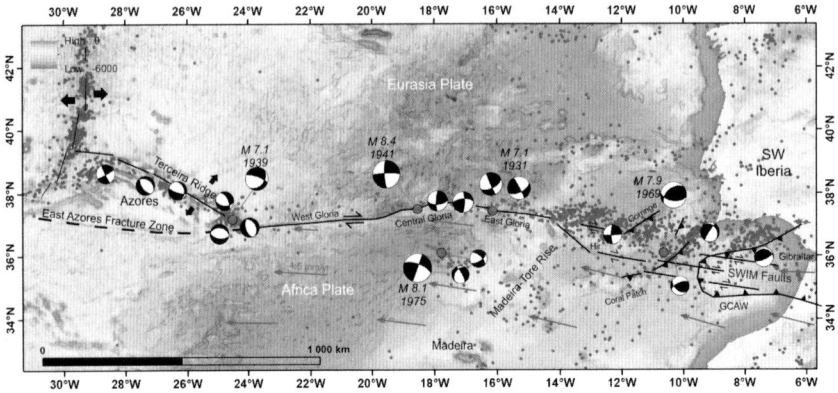

Fig. 13-7: The Azores-Gibraltar Fracture Zone. At Gibraltar itself, the pattern becomes more complex, with a fault-array rather than a single fault. Image: Hensen et al. (2019).

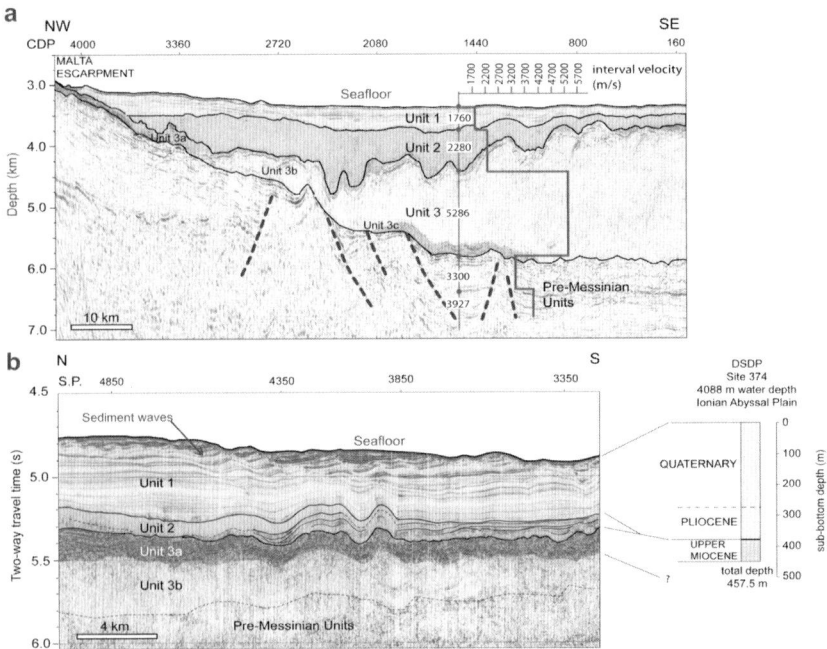

Fig. 13-8: Top diagram shows sediment wedges immediately against the east side of the Malta-Sicily plateau as seen in fig. 13-6. As can be seen in the lower diagram, there are three (or four) sedimentary units here again, overlying the pre-Messinian rocks. Note that the authors have spread these units out over millions of years from the Miocene to the present. Image: Micallef et al. 2018.

Since no flood of any erosive or transporting power can form merely by the slow overflowing of the sill by seawater, the separation of Africa from Europe, and the faulting at the Straits must have happened suddenly, and extremely rapidly, leading to a catastrophic flood that turned the freshwater Mediterranean back into saltwater, as various writers have said, and as we saw earlier.

In the central Mediterranean lies the Malta-Sicily plateau and figs. 13-6, 13-8 show that the flood overran the plateau and dumped enormous quantities of sediments on the eastern side, and against the Malta Escarpment. The seismic sections are similar to those shown earlier for the Aegean basins, and such an enormous quantity of sediment, draped as it is against the escarpment bespeaks massive amounts of water sweeping over the plateau. The "sediment waves" are obviously giant ripples such as are seen along the route of the Altai Floods; see fig. 13-1.

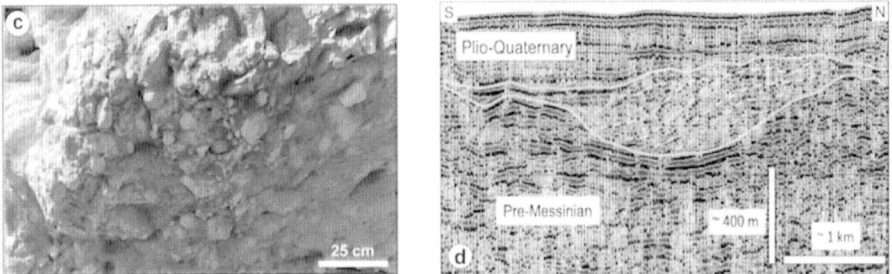

Fig. 13-9: On the left (c) is shown what are identified as Zanclean flood sediments from Sicily. They are described as: ". . . terrestrial chaotic Zanclean deposits." These deposits are found on land in south-eastern Sicily in the map section labelled (a) in fig. 13-6, and are thus similar to the deposits found offshore and shown in the above seismic cross-sections. On the right (d) is a seismic profile of a channel eroded in the "basal limestone" and infilled by flood deposits, the lowest of which is, no doubt, a "basal conglomerate" similar to the material shown in (c). Image: Micallef et al., 2018.

It is obvious from the map of fig. 13-4 that the Malta Plateau, lying between Sicily and Tunisia, is the gateway to the eastern Mediterranean and the natural route for the flood from the west. Given that nothing appears to be overlying the Zanclean sediments, which show "sediment waves" right at the sea floor in fig. 13-8, bottom diagram, and "sediment waves" are just giant ripples (fig. 13-1) produced by a flood, it would seem they were the last materials to be deposited. And since, like much else that lies on top of everything else, it must be relatively recent, I would have to consider it as much younger than 5.3 million years.

Finally, fig. 13-9 shows a fairly typical, poorly sorted, flood deposit of sub-rounded to rounded clasts in a muddy/mixed matrix, such as we've seen when examining the Ice Age in Greece. Judging by the scale provided, the largest clasts appear to be about 6-8 inches (15-20 cm) in diameter, with the majority being about 3 in. (7-8cm). The seismic profile adjacent shows a channel eroded into the underlying solid limestone, which can only be done by flood-driven, debris-laden water, and hence the logical conclusion is that the flood eroded the channel as it swept through, and filled up that channel as it slowed down, no doubt depositing a "basal conglomerate" while it was at it.

In conclusion, therefore, the evidence does indeed seem to indicate a major flood from the west not too long ago, and the fresh or brackish water also tends to confirm Hesiod and Pliny.

In agreement with orthodox geology, I consider there to have been major flooding and other disturbances at the end of the Ice Age; it's just that I don't consider the flooding as having anything to do with ice melting, as we'll see later, and I don't accept the ice-dammed lake theory of the Altai Flood. In fact, most theories of "ice-dammed lakes" and "outburst floods" do not stand up to scrutiny, as they do not appear to meet the pre-requirements of atmospheric physics or astrophysics, as I will demonstrate in due course.

Suffice it to say for now that, in the northern hemisphere at least, considering the air temperature is always warmer to the south, and it gets cooler moving north, ice, therefore, cannot exist south of liquid water. Ice and snow in the polar regions are melted by warm air blowing up from the south, so the ice and snow to the south will always be melted first. Hence, no so-called "ice dam" can build up to the south while ice and snow further north can receive enough heat to melt. The idea that sunlight can melt ice, on its own, is a false notion one often comes across.

Ice and snow are bright white, and in sunshine, dazzlingly so. Hence, we know, obviously because we're being dazzled, that the snow is reflecting most of the sunlight and absorbing very little. It does absorb some as the micro-surface of snow is uneven but all I ever see it do is cause some surface recrystallization or minor melting to produce an icy crust on the surface or else a granular layer.

Allied to the fact that bright as the sunshine may be in the clear dry air of the northern regions in winter, the sunlight is weak because it is impinging at a low angle, and therefore has very little heat to begin with. Anyone who lives in snowy regions will tell you that it doesn't matter if the sun is blazing out of an azure-blue sky all day long, the snow won't melt if the air temperature is below $0°$ C. At the same time, they'll also tell you that snow will melt all night long if the air temperature is any few degrees at all above $0°$ C.

Ice dams can only form in rivers that flow *north* in polar regions, such as those in Siberia, the Yenisei, Lena or Irtysh, as examples, and ice dams form on these rivers essentially every year. They occur because further south it warms up earlier, and the ice on the rivers melts while the river is still frozen in the north. The river then brings ice rafts, floes, and slush north to jam up and dam the rivers, causing floods—or they would if the Russian air force didn't bomb the dams out of existence every year.

It's fairly clear that some large-scale event occurred about 5,000 years ago, as we saw from the evidence on Greece, and that "event" appears to have been the last, or second-last, major occurrence to affect Greece. While it may have consisted of a large-scale flood from the west, I'm inclined to think that that flood, from the west, is somewhat older than a mere 5,000 years. Whenever this flood did actually occur, I have to conclude that it is what has been mistaken for the 5.3-million-year-old Zanclean Flood.

The flood from the north is hazarded (by me) to have occurred at the end of the Ice Age, at the time of Atlantis's destruction. However, I may be wrong about which is which, and the one from the west might be the one related to Atlantis, while the one from the north may be an even older flood, given the limited study this essay represents. Obviously, much more work needs to be done, though the flood from the north does appear to be a definite earlier one.

When I say "second-last," I'm here referring to the Exodus "event" which may have been described in the Ipuwer Papyrus, and which would be the last major out-of-the-ordinary occurrence, as far as I can tell. I do not at present intend to address this issue.

We will now move on to the "legendary floods of Greece," which, like the "legends" of the Makah and Klallam, and countless others, are quite possibly a good deal more valid than modern science says they are.

Chapter References

Abril J. M. & Periáñez R., 2016, Revisiting the time scale and size of the Zanclean flood of the Mediterranean (5.33 Ma) from CFD simulations, *Marine Geology*, Vol. 382, Dec., pp 242–256.

Baker, V. R. et al., 1993, Paleohydrology of Late Pleistocene Superflooding, Altay Mountains, Siberia, *Science*, Vol. 259, Jan. 15, 1993.

Carnevale, G. et al., 2018, Fossil marine fishes and the 'Lago-Mare' event: Has the Mediterranean ever transformed into a brackish lake? *Newsletters on Stratigraphy*, Vol. 52, No. 1, pp 57-72.

Caruso, A. et al., 2020, The late Messinian "Lago-Mare" event and the Zanclean Reflooding in the Mediterranean Sea: New insights from the Cuevas del Almanzora section (Vera Basin, South-Eastern Spain), *Earth-Science Reviews*, Vol. 200, 102993.

Garcia-Castellanos, D. et al., 2020, The Zanclean megaflood of the Mediterranean—Searching for independent evidence, *Earth Science Reviews*, Vol. 201, Feb., 103061.

Gargani, J. et al., 2008, Evaporite accumulation during the Messinian Salinity Crisis: The Suez Rift case, *Geophysical research Letters*, Vol. 35, Iss. 2.

Grossi, F. et al., 2008, Late Messinian *Lago-Mare* ostracods and palaeoenvironments of the central and eastern Mediterranean Basin, *Bolletina della Paleontologica Italiana*, Vol. 47, No. 2, pp 131-146.

Hensen, C., 2019, Marine Transform Faults and Fracture Zones: A Joint Perspective Integrating Seismicity, Fluid Flow and Life, *Front. Earth Sci.*, March 2019 *Sec. Structural Geology and Tectonics*, Vol. 7.

Hesiod, *Theogony*.

Hsü, K, 1983, *The Mediterranean Was a Desert*, Princeton University Press, Princeton New Jersey.

La Violette, P., 2017, The Generation of Mega Glacial Meltwater Floods and Their Geologic Impact, *Hydrology Current Research*, Vol. 8, Issue 1.

Loget, N. et al., 2006, Mesoscale fluvial erosion parameters deduced from modeling the Mediterranean Sea level drop during the Messinian (Late Miocene), *Journal of Geophysical Research Atmospheres* 111 (F3): 1-15.

Micallef, A., et al., 2018, Evidence of the Zanclean megaflood in the eastern Mediterranean Basin, *Scientific Reports*, Article number: 1078 (2018).

Nemec, W. & Postma, G., 1993, Quaternary alluvial fans in south-western Crete: sedimentation process and geomorphic evolution. *Special Publication of the International Association of Sedimentology*, Vol. 17, pp 256-276, cited in Hughes et al., 2006.

Pliny, *Historia Naturalis*.

Roveri, M. et al., 2014, The Messinian Salinity Crisis: Past and future a great challenge for marine scientists, *Marine Geology*, Vol. 352, pp 25-58.

Rudoy, A. N., 2002, Glacier-dammed lakes and geological work of glacial superfloods in the Late Pleistocene, Southern Siberia, Altai Mountains, *Quaternary International*, 87, 119-140.

Runnels, C. et al., 2014, Paleolithic research at Mochlos, Crete: new evidence for Pleistocene maritime activity in the Aegean, *A Review of World Archaeology*, Vol. 88, Issue 342.

Ryan, W. et al., 1997, An abrupt drowning of the Black Sea shelf, *Marine Geology*, Vol. 138, No. 1-2.

Ryan, W. & Pitman, W., 1998, *Noah's Flood, The New Scientific Discoveries About The Event That Changed History*, Simon & Shuster, New York, Touchstone Ed. (2000).

Saner, B. R. M., 1928, River terraces and Raised Beaches: The Work of the Commission on Pliocene and Pleistocene Terraces at the International Geographical Congress, 1928, *Science Progress in the 20th Century (1919–1933)* Vol. 24, No. 95 (January 1930) Sage Publications Ltd.

Suess, E., 1904, *The Face Of The Earth, (Antlitz Der Erde)*.

Vai, G. Battista, 2016, Over half a century of Messinian salinity crisis, *Bolletin Geológico y Minero*, Vol. 127, No. 2/3, pp 625-641.

Vitaliano, D, 2007, Geomythology: Geological origins of myths and legends, *Geol. Soc. Lon. Special Publications* No. 273, pp 1-7.

CHAPTER FOURTEEN:

LEGENDARY FLOODS
OF GREECE

THE ANCIENT GREEKS THEMSELVES HAD THREE floods; that of Ogyges, Deucalion, and Dardanus. In which case, according to the priest, there are two more floods unaccounted for, prior to that of Deucalion. The flood from the south, over Crete, may have been one of these, resulting from some disturbance or other in the Mediterranean/North African region. We recall the alluviation event on Greece about 5,000 years ago. You can't have alluviation without the water necessary to do the depositing that would constitute that "event."

As to which of the named floods, if any of them, that this alluviation event can be ascribed to, is impossible to say at this point, as one can easily imagine. Hence the need (or understandable desire on my part) for Greek geologists, preferably, to work on the question.

Given my extreme skepticism, or to be more accurate about it, downright rejection, of the chronometrical abilities of Charles Lyell's backside, and of orthodox geology's thence-derived timescale, I consider it more than possible, and highly likely, that the Zanclean Flood did not occur 5.3 million years ago. There is plenty of geological evidence that a major flood came through the Pillars of Heracles not too long ago and swept over part of northern Africa-Sahara, as we'll see later from Spratt and his travels, yet again. We might, as a matter of interest, also be able to account for the salt in the Sahara.

Further geological evidence comes from the general Mediterranean region, as does a certain amount of archaeological evidence, particularly from the Maltese islands, which exhibit signs of having suffered some major disturbance involving flooding as well as fracturing and collapse. Since ancient building complexes on Malta and Gozo are all affected in a similar way, it would appear that some event affected the whole region at one time not too long ago.

For example, many giant stone blocks, weighing upwards of 30 tons apiece, and forming the east and west walls of these building complexes, are all thrown down to the east, as though they were hit by a major force coming from the west. The north and south walls are relatively undamaged and still standing. This means, of course, that this event had to have been recent, after the buildings were built, obviously.

The flood of Ogyges is considered the earliest of the three, and was thought by some to have been worldwide, but by others to have been just a local flood, which is the usual assumption, and affected only the region of Attica, i.e., the Athens area, or Boeotia in the form of the flooding of Lake Copais.

Therefore, this flood can be considered, as many do, that it refers to the general area of Lake Copais down to the region of Attica. Looking at the map in fig. 14-1, it seems difficult to locate a source for a large quantity of water, such that this flood could be considered as more than a run of the mill tsunami. Still, the logic we discussed in volume 1, where only the out-of-the-ordinary is myth-worthy, an ordinary tsunami or outburst flood, or anything relatively normal would not make the grade. Hence, it must have been abnormal.

Fig. 14-1: Map showing Lake Copais and the Attica area. Lake Copais is 91 m (300 ft.) above sea level, so a tsunami seems unlikely. Image: Wikimedia Commons, CC-BY-SA-3.0 Lic Photo: georg-hessen.

The only other suggestion I would make vis-à-vis the flood of Ogyges is that it may relate to the island of Ogygia out in the Atlantic, as Plutarch described. Hence this may be a flood from the Atlantic and possibly at the time of Atlantis' destruction.

Given the name "Dardanus" and the straits called the "Dardanelles" at the western end of the Sea of Marmara, I consider it probable that the Flood of Dardanus refers to the Altai Floods, which supposedly resulted from the failure of ice-dammed lakes, the favorite explanation for all evidence of massive flooding that dates to the end of the Ice Age.

At the same time, there is a severe shortage of evidence for the presence of ice on the Altai Mountains during the Ice Age. As a general statement, there is no evidence whatever that Siberia was, in any way, ever affected by ice. Despite what Ice Age maps show as the extent of ice cover, the reality is that no evidence for ice has ever been found east of the Barents Sea, which is very far west, and this lack of evidence extends all the way across Russia, Asia, and Alaska, and across north America, and eastward almost to Hudson Bay, Canada, as shown by the lines and arrows on the map of fig. 14 2.

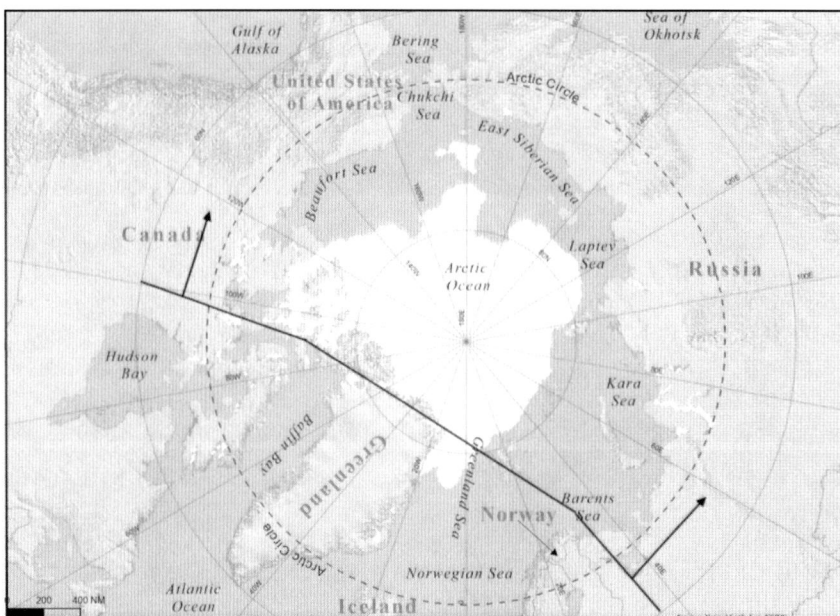

Fig. 14-2: Map of the Arctic Ocean and circum-polar region as it is today. Despite what geology says, and as we'll see in great detail later, no evidence exists that ice was ever present, during the Ice Age, in the area indicated by the lines and arrows. Greenland is presently, of course, covered in ice, so we don't know whether there was ice on it during the Ice Age or not. However, if we recall, there were swamp cypress growing on it at the time, so the presence of ice is questionable. Also, the Altai Floods must clearly have a different explanation. Image: "Law of the Sea" Policy Primer, Fletcher School, Tufts University.

While many a reader might be taken somewhat aback at the foregoing statement, given the standard story of an American continent half-covered in ice "a mile high" as they always say, and from one shining sea to the other, the fact is that much of what passes for evidence here does not even come close to satisfying R. F. Flint's criteria for ice diagnosis. There are also quite a few issues with regard to atmospheric physics and astrophysics that call these claims of a continent half-covered in ice into serious question. We will, of course, examine all this very thoroughly when we finally get around to it, but we haven't forgotten all those unmolested Pliocene gravel deposits with their shells and plant fragments perfectly preserved.

A large flood coming west through the Dardanelles would account for the sediment deposits just west of the Dardanelles Straits. Whatever the origin of this flood, it is considered to have been among the largest in the world, if not the largest, and it came all the way from Siberia/China. To make it all the way to the Aegean would certainly imply that an enormous amount of water was involved.

It so happens that the Altai Mountains are not far away from the Tarim Basin, wherein now lies the Taklamakan Desert, which was once a large inland sea or lake, known as the Turan Sea, and was part of the Paratethys, forming its eastern end (Zheng et al., 2023). To the north, as an extension, was the Siberian Sea. Both seas surrounded the Altai Mountains on their west side, as shown in fig. 14-3. Both of these seas emptied at some point in the past, and we recall the writers in chapter 13 that referred to freshwater from the Paratethys entering the Mediterranean in the form of a "dump," which obviously suggests a large volume all of a sudden, i.e., a large flood.

While the Tarim Basin and its general region are arid today, the environs of the basin were well-populated thousands of years ago, and by strangely tall (up to 6' 6" [2 m]) people with both fair and red hair and blue eyes, quite unlike the locals today. They have been compared with the Celts of western Europe, and they did indeed wear tartan clothing and used implements and weapons typical of Celts, while one woman was considered somewhat Nordic in appearance.

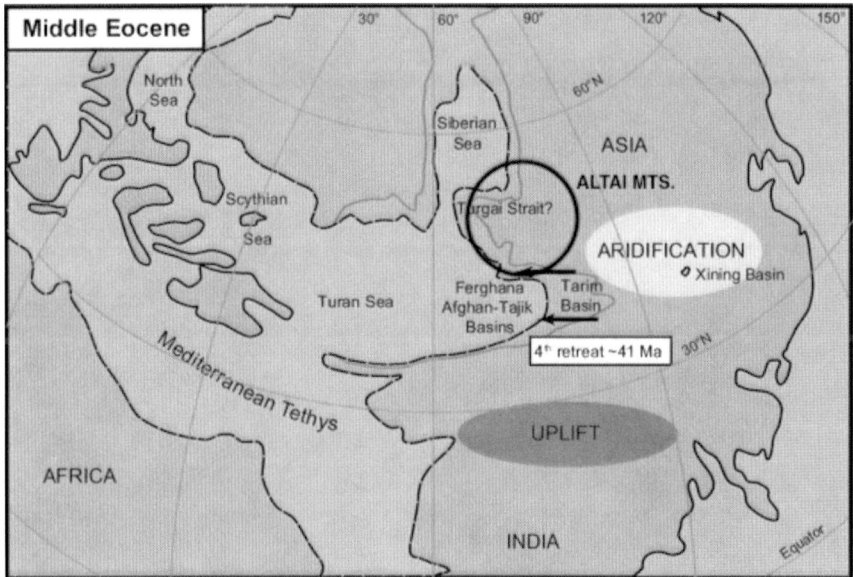

Fig. 14-3: Paleogeographic map of the Paratethys, Turan and Siberians Seas. The Altai Mts. are directly south of the Siberian Sea, and hence the Siberian and Turan Seas, rather than the Altai Mts. may have been the source of water for the Flood of Dardanus. Image: Zheng et al., 2023.

Their dead were buried in the dry salty soil, which caused mummification, and these mummies can be seen today in a museum at Ürümchi, the capital of the Xinjiang Uyghur Autonomous Region of China. Rather than investigate the question seriously, certain of our more conservative academic institutions have declared these people to be genetically isolated and just descendants of some local (spurious) Ice Age survivors. A priori dismissal hardly seems the appropriate treatment of what appears to be a serious human question here.

As there are certain political sensitivities surrounding Xinjiang at present, the mystery of the Tarim mummies will have to remain unresolved for now. Since an investigation of the whole Tarim Basin question would require a book of its own, and one best written by a Chinese scientist, for the time being we will not concern ourselves too much with the origins of the Tarim Celts, or their mummies, and move on to the Flood of Deucalion.

The Flood of Deucalion is the only one mentioned by Solon, and it was apparently known to the Egyptian priest. This flood legend bears some similarities to the Noachian one, in that Deucalion was warned beforehand, and he and his wife Pyrrha survived in a "chest" which came to rest on a mountain.

Different versions exist, involving pairs of animals, or not, and different mountains on which the chest, or ark, came to rest. Rain is also involved, and one version says that the flood lasted nine days. Oddly enough, this is about the length of time the flood of the Makah and Klallam, of the Pacific Northwest, lasted, with the water rising four days and ebbing four more. As far as these flood legends go, we will recall the comparison with Noah's Flood mentioned by Swan (1868).

However, Deucalion's flood has been dated to 1539 BCE according to Vitaliano (1973, p. 158) said date being derived from a pillar on the island of Paros which lists Deucalion as a king. This time period was also suggested by Galanopoulos, which would put it around the time of the Exodus, as well as the eruption of Santorini. The interpretation, therefore, is that a tsunami was responsible (Vitaliano, 1973) an idea I would not accept for the many reasons given in volume 1, as well as herein.

Instead, because of the similarity to the Noah and Mesopotamian legends, I would consider a dating in the same range of 5,000 years as not too far beyond the bounds of believability, in terms of it still being within the folklore memory of ancient Greeks such as Solon, in the same way as was the eruption and collapse of the giant volcano of the Klamath, leaving the Three Sisters. At the same time, and since we're dealing with legend, this flood may have been an earlier one, such as that of Ogyges, for instance.

Thus, this flood might obviously coincide with van Andel and Zangger's (and others) "alluviation events" of about this time, and provide the needed water. Further, these writers described and discussed a major alluviation event, but little or no evidence of extreme violence was presented. Hence it may have been a more gentle flood, which, by the sounds of it, may have been the same relatively gentle event as reported by the Makah and the Klallam which we discussed in volume 1.

None of these three floods, however, even *hint* at the scale of the catastrophe that would be required to sink a continent, and the enormous disturbance of the Atlantic occasioned by such an occurrence would be such as to set that entire ocean in motion. There can be no question but this motion would resemble that of an oceanic disturbance developing into a tsunami, which are caused by similar means, but it would be of such an enormously large scale as to beggar description, but the tsunami "analogy" does hark back to Lyell's complaint vis-à-vis the catastrophists and their lack of "reference to existing analogies."

However, as the disturbance approached the shore, the water would build up into a giant wave, up to many hundreds of feet or meters deep, or more, that would sweep in over the land at enormous speeds, tear it and everything on it to shreds, and sweep it all away to dump it somewhere as the wave's power declines. The wave would, in fact, be suggestive of Capt. Spratt's "wave of translation" or James Hall's "great diluvial wave" sweeping across the country, as in the analogy of an enormous tsunami.

Chapter References

Cooan, C., 2006, A meeting of civilizations: The mystery of China's Celtic mummies, *The Independent*, U.S. Edition, August 8.

Vitaliano, Dorothy, 1973, *Legends of the Earth: Their Geologic Origins*, Indiana University Press, Bloomington / London.

Zheng, H. et al., 2023, Birth of the Taklamakan Desert: When and How? *STEM Education*, Vol. 3 (1): 57-69.

CONCLUSION

THE FOREGOING CHAPTERS PROVIDED MANY OF MY reasons for viewing the orthodox explanations for the present-day condition of Greece with some skepticism. There appears to have been far too much water involved, and no known source, since Greece, supposedly at least, has been as dry as it is today for the past 12,000 years, and cold, and likely even dryer, during the so-called Ice Age, for any number of thousands of years before that.

This is generally what the various writers we've reviewed have said, and this picture was modified or interrupted by various, generally unexplained, fluctuations and events over that time. The old priest, of course, had a very simple explanation for the condition of Greece. We have seen what that explanation was, or is, and we have also seen what the multitude of orthodox so-called explanations are. And I have attempted herein to make the case that those orthodox explanations just do not cut the mustard when it comes to accounting for the evidence on Greece. Violence and catastrophe were obviously, and heavily, involved, because the laws of physics say they were, and common sense tends to agree.

We can conclude, therefore, that the priest's deluge was a major part of the catastrophe that destroyed Atlantis and that it was very widespread. It overran what was then Greece and massively eroded the surface of the land, washing everything loose into the sea and into all of those subsiding basins. It follows that if the catastrophe drowned Atlantis in the middle of the Atlantic and overran Greece, then it must have affected everywhere in between, including areas considerably further to the east than Greece, to a lesser degree, presumably.

By the same token, it's just as reasonable to suppose it seriously affected areas as far to the west of the middle of the Atlantic, as Greece is to the east of it. This would take us well into North America, and it would imply that part, at least, of the north American continent was also overrun by an enormous deluge.

If so, it shouldn't be all that hard to find evidence of large-scale flooding extending across large areas of the land surface of the northern European and American continents. And, since it's difficult to see how such a large flood, involving a highly disturbed ocean, could be confined to the northern hemisphere only, it is reasonably safe to conclude that it also affected the southern and eastern hemispheres.

This means that we are talking about a global flood that affected different areas of the planet to a greater or lesser extent, and one that can only have occurred as a part of a general global catastrophe. Given that there seems to be no agent capable of acting from within the planet, this catastrophe could only have been induced by some cosmic influence or extraterrestrial agent.

Considering the prevalence of catastrophe legends all over the world, it is clear that many people survived the event in different places, hence the notion of the lone pair of survivors seems to be simple hyperbole.

However, regarding any major floods we have experienced in the modern world, or have any historical knowledge of, the numbers of people killed has never been so high as to warrant use of the term "a destruction of mankind" as the priest said. Hence, the "destructions" we must be dealing with here are catastrophes far outside the bounds of our historical experience, our uniformitarian geological theories, or perhaps even the limits of our imaginations.

At the same time, however, it is obvious that none of these catastrophes resulted in the total annihilation of the world's human or animal populations.

It is really not to be expected that a global catastrophe could affect all parts of the globe equally, considering the uneven distribution of land and sea. Also, it's more than likely that the effect of any outside agent would be focused more in some area than others. After all, the priest made a distinction between the catastrophe by water and that by fire, if only by the difference in effect between highlands and lowlands, and that in the context of survivors.

Hence, we can be fairly sure that if a warning were possible and provided, then such disasters would likely be reasonably survivable, with adequate preparation. Needless to say, such a warning would depend on knowing as much as possible about the last one, which is, in a preliminary way, essentially what I am attempting here.

However, notwithstanding the general anti-catastrophist stance of the geological establishment, if large-scale catastrophes have occurred, then their signatures should be evident in the record of the rocks, and the recent sediments visible and accessible at, on, and under the surfaces of the earth and the oceans. Such signatures should consist of clear and distinct breaks in process continuity, and sharp changeovers to distinctly different conditions, all of which we saw in Greece. An obvious second line of enquiry is the Archaeological record and whether it too shows disruptions and breaks in continuity.

Our detailed, but by no means exhaustive, survey of the geology of Greece seems to be at distinct odds with the principles of uniformitarian geology. We have seen that the many geologists who have studied Greece persistently encountered phenomena that defied simple and direct explanation. We have seen that the notion of ongoing everyday causes being capable of doing the work seen done fails practically every test, and little evidence of their effectiveness was presented.

As we saw in Greece, many geologists played, or had to play, rather fast and loose with the laws of physics in their efforts to obey Lyell's rules of uniformitarianism. While we did not get into any mathematical expressions of the physical laws pertaining to the many phenomena we examined, or do any mathematical analyses, we all know well enough that large results require large forces, e.g., large boulder transport.

Many phenomena we saw obviously cannot be produced slowly and gradually, e.g., gravel and conglomerate deposits, and specific physical prerequisites are necessary to produce others, e.g., water or ice must be carrying abrasive materials to effect hard rock erosion. We will meet all these questions again, as well as many other similar problems, and we will analyze them more rigorously at the appropriate time.

However, since volume 1 and 2 address the Atlantis legend in particular, and hence are a two-volume set, and since the whole question was whether uniformitarianism was as valid as it's claimed to be, I believe I have gone some way to showing that it has difficulties when it comes to explaining many phenomena, in Greece and in general.

And, since most of the geologists consulted herein averred that the Ice Age, and related agents, were the main causes of the condition of the Greece of today, in their view then, uniformitarian processes are merely acting on this landscape *pre*-formed by that Ice Age.

I, on the other hand, have to conclude from the same evidence that the Ice Age had little to no effect on Greece, and present-day processes are merely acting on a landscape *pre*-formed by one or more catastrophic agents, which I do not think were in any way related to an Ice Age.

Having now a general picture of the recent geology of Greece and the Aegean, and despite what orthodox geology says, we have seen that there is a great deal of evidence that agrees very well indeed with the narrative of that old Egyptian priest.

And, speaking of priests, there is one other flood legend I heard, from one or another Irish priest, when I was far too young to know better than to listen to them. That is, of course, the Flood of Noah and its possible parentage in the Sumerian Flood myths, i.e., those known variously as the Flood of Utnapishtim, Ziusudra, or Xisuthrus, and all considered mere local floods, as is usual.

Since Mesopotamia is just over the road from Greece and the Mediterranean, any evidence to be found there might provide some further enlightenment regarding the occurrence of these enormous floods of legend. As with the floods of Greece, we will examine Mesopotamia from the point of view of geology and archaeology only, merely referencing any mythology for context.

The next and third volume, therefore, is something of a stand-alone insert into the main Atlantis narrative, but quite relevant nonetheless, as it may reveal much valuable evidence.

We will then turn to ancient archaeology, which potentially has answers to assist in a few lines of inquiry, which would be in the areas of catastrophic destruction and/or submersion of buildings and cities, etc., breaks in continuity, and lastly, the existence or otherwise of advanced, shall we say, anomalous, ancient civilizations.

That is not to say that we are seeking, or expecting to find, jet planes or helicopters in Egyptian temples, but rather things that just don't really "fit." That is, things that don't fit the picture of the ancient past that we've been fed, and consisting of what are dismissively called "anomalies" or "out-of-place-artifacts" (ooparts) of one kind or another. These usually represent knowledge, abilities, and technologies we would not expect the people of that period thousands of years ago to possess, at least based on what our orthodox archaeologists have told us.

Since I've demonstrated in this volume that Greece, for one, does not seem to obey the British "rules of the game," it is really for Greek geologists to assess my claims, if any have a mind to. Ireland, mind you, was never inclined to obey any British rules either, as its history attests, so we'll see what the Irish story is in due course, since the country plays a rather significant, but perhaps little known, role in both the development of geology and the problem of Atlantis.

In the following volumes, therefore, we are going to look at various archaeological sites and other phenomena that seem to call into question the standard primitive-to-sophisticated chain of history. We will explore Mesopotamia in volume 3, and then, in volume 4, take a tour of the eastern Mediterranean, from Greece around to Egypt, on to Malta and Sicily, and thence to the Pillars of Heracles, and finally out into the Atlantic, where Atlantis lies sleeping in the depths, waiting for us, as easy to find as it has always been. It's recognizing it when we do that's the hard part.

However, before we do move on, it is obvious that the foregoing overview of the geology and paleontology of Greece is far from highly detailed and comprehensive, and I do not claim to be perfectly correct in everything I said. The objective originally was to analyze the evidence on Greece and assess whether academic geology's uniformitarian explanations were the correct ones, or whether there may be other possible interpretations.

I have to conclude that there are, as I have attempted to show. I have also claimed repeatedly herein that the doctrinaire explanations of orthodox geology are utterly inadequate to account for the evidence, and I do claim to have made a reasonably good case for that assertion, despite any errors that are all my own.

Since I have made clear my differences with British geology, and since at least some Greek geologists were willing to contemplate some catastrophic action in the past, I would hope that some may follow this work and assess it against the evidence on Greece, evidence with which they are, of course, much more familiar than I.

Further, as shown herein, much of that evidence is just lying on the surface—on the hills, in the basins, along the river valleys, and at the coasts, ready and waiting, not only for Greek geologists, but for the people of Greece to freely examine at their leisure.

In light of my claims as to uniformitarianism's inability to explain the geological history of Greece, the opinion of Greek geologists would, of course, be of the greatest interest. Should any agree with any claims made herein, then they may also agree that Greece has had a very different geological history, a history that, to me, is clearly marked by Greek catastrophe and not British uniformity.

As for Plato's Atlantis Legend, Greek geologists, and indeed, the Greek people, are in a perfect position to check my claims, and decide whether the evidence on Greece, merely sampled and exampled herein, supports the eloquent words of their own ancient philosophers—or the incoherent ramblings of an 18th century Scottish farmer.